11.95

GENERAL RELATIVITY FROM A TO B

GENERAL RELATIVITY FROM A TO B

Robert Geroch

The University of Chicago Press

CHICAGO & LONDON

The University of Chicago Press, Chicago 60637
The University of Chicago Press, Ltd., London

Printed in the United States of America
88 87 86 85 84 83 82 81   9 8 7 6 5 4

Library of Congress Cataloging in Publication Data

Geroch, Robert.
General relativity from A to B.

Includes index.
1. General relativity (Physics) I. Title.
QC173.6.G47   530.1'1   77-18908
ISBN 0-226-28863-3
0-226-28864-1 (paper)

Contents

Preface

This book grew out of the lecture notes for a course I gave, in 1975, to nonscience undergraduates at the University of Chicago. It discusses an area of physics, but it is also intended as a statement of the sort of material in science which I feel those outside this field would find most useful and most interesting: not definitions and the substitution of numbers in formulas—this is not even physics, not levers and inclined planes—the physics of the eighteenth century, and certainly not a view from below, with choreographed "gee-whizes," of a tower shrouded in mystery. Rather, it is my hope that there is here a glimpse of how physics actually works, and how working physicists think about the world in which we live.

Introduction

Einstein's general theory of relativity is a theory of gravitation which (for reasons perhaps not altogether understood) also touches on the structure of space and time. It is easily the most widely accepted theory of gravitation available today; its study is an active and growing branch of modern theoretical physics. It is our purpose to understand what this theory is—how it works, what it has to say, what the physical phenomena are with which it deals, and what its predictions and applications are.

Why is a new, fancy theory of gravitation necessary? What is wrong with the old theory—Newton's law of gravitation—which appears to describe gravitation both simply and accurately with a single law: There is a force between any two bodies in the universe, proportional to their masses and inversely proportional to the square of the distance between them. A complete answer to this question would be long and technical; we here only sketch the idea. At the beginning of this century there was available, in addition to Newton's theory of gravitation, a theory, due to Maxwell, of electric and magnetic effects. The difficulty that was ultimately to lead to the downfall of Newtonian gravitation arose originally as a difficulty associated with Maxwell's theory. The latter suffered from the following, unfortunate, affliction. Consider some system whose elements interact with each other through electric and magnetic forces. Let a person stand beside this system observing it, and let us suppose that the system is correctly described, according to this observer, by Maxwell's theory. Now let this same system also be observed by a person walking by it at constant speed. We ask, Will this second observer necessarily also see this system as satisfying the precepts of Maxwell's theory? This is really two questions, an experimental one (What will happen if the experiment above is actually performed?) and a theoretical one (Ignoring experiments, and just assuming that the system satisfies Maxwell's theory according to the first observer, will it according to the second?). It turned out that the answer to the theoretical question was no (although the discrepancies, according to the moving observer, from Maxwell's theory would be too small to be seen in an actual experiment with the then-present technology). Maxwell's theory, it appeared, was not "universally valid."

What (so it was asked at the time) is to be done about this problem? The obvious answer would be to attempt to somehow modify Maxwell's theory so as to avoid this malady, and numerous attempts were made along this line. The ultimate resolution, however, turned out to be along quite different lines. Instead of modifying Maxwell's theory, one modifies the assumptions, implicit in the argument above, concerning the relationship between observations made on a single system by two persons moving with respect to each other. By choosing the latter "modification" appropriately, one is able to retain Maxwell's theory in its original form and at the same time remove the difficulty. Unfortunately, a new problem is now created. As a consequence of this "modification of the assumptions concerning the relationship between observations made on a single system by two persons moving with respect to each other," Newtonian gravitation now succumbs to the malady originally present in Maxwell's theory. (That is to say, repeating the thought experiment above, but now for a gravitating system and now using the new "assumptions concerning the relationship . . . ," the moving observer now does not see the system satisfying Newton's law of gravitation!) Thus, if we accept the resolution offered for the original problem with Maxwell's theory, we create a problem for Newtonian gravitation. This new problem, in turn, was resolved through a new theory of gravitation—Einstein's theory.

It is perhaps not too great an oversimplification to say that there have been two great revolutions in physics in our century: quantum mechanics (perhaps the more radical of the two) and general relativity. The latter qualifies as a revolution because it forced us to a view of nature that was significantly different from what we had before. It had been thought previously that the "assumptions concerning the relationship . . . " were obvious and immutable, subject to neither debate nor change.

Lack of time (for the treatment of both the theory itself and its mathematical underpinnings) prevents us from treating the theory in its full technical detail. Fortunately, its basic structure and fundamental ideas are near the surface. It is by no means difficult to understand what the revolution was all about, and this is what we wish to do.

A

The Space-Time Viewpoint

1

Events and Space-Time: The Basic Building Blocks

The notion of an event is the basic building block of the theory. It will dominate all that follows.

By an event we mean an idealized occurrence in the physical world having extension in neither space nor time. For example, "the explosion of a firecracker" or "the snapping of one's fingers" would represent an event. (By contrast, "a particle" would not represent an event, for it has "extension in time"; "a long piece of rope" has "extension in space.") By "occurrence in the physical world" we mean that an event is to be regarded as a part of the world in which we live, not as a construct in some theory. Of course, there are many events around: some occurred long ago, some are occurring now, and others will (presumably) occur in the future. What is meant by "idealized . . . having extension in neither space nor time" requires more explanation. Consider the explosion of a firecracker. The explosion lasts for some finite time (say, one-tenth of a second), and so this occurrence has extension in time; the explosion takes place over some finite region of space (say, one-quarter of an inch), so it has extension in space. If, however, we used a smaller and faster-burning firecracker, these "extensions" would be smaller. An event is to be an idealization of this situation in the limit of a "very small, very fast-burning" firecracker. (The situation is similar to that which would arise from the statement: "A point on the blackboard is an idealized chalk mark having extension neither up-down nor right-left." This analogy goes a little deeper: events will shortly become "points" of an appropriate space.)

We regard two events as being "the same" if they coincide, that is, if they "occur at the same place at the same time." That is to say, we are not now concerned with how an event is marked—by firecracker or finger-snap—but only with the thing itself.

Is one to regard events as "existing" even if there is nobody there to mark them with finger-snap or otherwise (for example, in a dark, empty closet at 3 A.M.)? It is part of what we wish to mean by an event that the answer is to be yes. Perhaps it would have been better to say originally "An event is an idealized *potential* occurrence. . . ." As a general rule, failure in physics to attribute "existence" to things not directly perceived leads to various difficulties of the "If a tree falls in the forest and nobody is there to hear it, does it still make a sound?" variety. Failure to do so in the present case would, as far as I can see, make further development of the theory virtually impossible. This is not to say that such questions are uninteresting or unimportant. Rather, it has become the custom in physics to relegate them to others by the practice of being liberal in bestowing "existence."

Are events real? What are they really like? These questions are dealt with (more accurately, avoided) by means of another custom. Physics does not, at least in my opinion, deal with what is "real" or with what something is "really like." The reason, I suppose, is some combination of (1) One does not know how to effectively attack such questions. (2) One does not know what sort of thing would represent an answer. (3) These questions are too hard. In any case, one conventionally deals with relationships between things which one does not (or perhaps cannot) understand on a deeper level. One does, of course, sometimes come to understand some basic concept more deeply. (For example, space and time were basic concepts in Newtonian gravitation. With general relativity, one does feel a sense of deeper understanding.) Perhaps it is true to say that one has found from experience that deeper insight into the basic concepts of a theory comes most often, not from a frontal attack on those concepts, but rather from working upward into the theory itself.

Relationships between events—that is what we are after. Virtually everything we say hereafter can be resolved, directly or indirectly, into some statement of such a relationship.

We wish to discover the "correct" theory of the relationship between events. It is instructive to arrive at the final theory indirectly, through a sequence of preliminary attempts. We begin then with the rather naive view of everyday experience, a view which will subsequently be found to be inappropriate.

According to the Aristotelian view, an event is naturally characterized by giving its position in space together with the time of its occurrence.

We can make this view more explicit. Let there be set up, within a room, a Cartesian coordinate system x, y, z. That is to say, each position in space is to be described by three real numbers: the value of x, the value of y, and the value of z. For example, the "value of x" might be the distance of that position from one side wall, the "value of y" the distance from the front wall, and the "value of z" the distance from the floor. Our coordinate system permits, then, "numerical location of positions." The position described by $x = 12$, $y = 3$, $z = 9$ is that located 12 feet from the side wall, 3 feet from the front wall, and 9 feet from the floor. Now let the room be filled solidly—wall to wall, floor to ceiling—with people. Each person always maintains his same position within the room. Each person can describe his fixed position, then, by giving the appropriate values of x, y, and z. Let those values be printed on a small badge which each person wears. Next let there be distributed, to each of our subjects, an accurate watch. These watches are all synchronized (for example, by having another person communicate with each person and compare his watch with theirs).

Imagine, then, the arrangement sketched above. We use this arrangement to characterize events as follows. Let some event be chosen, marked, say, by the explosion of a firecracker. Since our subjects are packed solidly, one of them will be in the immediate vicinity of the explosion. Let that person write on a slip of paper the three numbers (x, y, and z values) which appear on his badge, and also a fourth number, the time, according to his watch, at which the explosion was experienced. This slip of paper is then passed forward to a moderator desirous of knowing our characterization of this particular event.

Of what does our characterization consist? Of four numbers, the values of x, y, z, and t (time). The first three numbers give the "position of the event in space"; the fourth gives the "time of its occurrence." We are here characterizing events, then, according to the Aristotelian view.

Why did we go on and on, taking the trouble to be so explicit and so careful about such a simple idea? There are several reasons. The characterization of physical phenomena (such as events) is supposed to be grounded on a more or less explicit set of instructions for actually carrying out the characterization experimentally in the physical world. Normally, it is pretty clear what is to be done, and the instructions need not be given in great detail. Here, however, our concern is the structure of space and time itself, and care in saying exactly what we mean is not an empty exercise. A more important reason is that, as we shall see later, implicit in the construction above are certain assumptions about the way space and time operate. These various assumptions, it will turn out, are simply not true in our world. It is convenient, therefore, to have the present characterization in sufficient detail that we can later pick out these assumptions.

We claimed earlier to be concerned with relationships between events, not "views." What relationships, then, are implied by the Aristotelian view? Let there be given two fixed events. We ask whether, in the Aristotelian view, each of the following makes sense.

Do the two events have the same position in space? Since an event is here characterized by giving its position in space together with the time of its occurrence, this question does make sense. In terms of the explicit formulation, let the first event be characterized by values x_o, y_o, z_o, t_o, and the second event by x_o', y_o', z_o', t_o'. We may say that these two events have the "same position" provided $x_o = x_o'$, $y_o = y_o'$, and $z_o = z_o'$.

Do the two events occur at the same time? Again the question makes sense—explicitly, we may say that the two events "occur at the same time" provided $t_o = t_o'$. (Clearly, two events are the same provided they have both the same position and the same time.)

What is the distance between the two events? This question also makes sense. Since an event is characterized by, among other things, its position in space, we can simply compute the "distance in space between those two positions." Explicitly, the computation would be as follows. We first compute $x_o - x_o'$, the difference between the x values of the two events; then $y_o - y_o'$, the difference between the y values; then $z_o - z_o'$. The distance between the two events would then be given, according to the Pythagorean theorem, by the following equation: $(\text{distance})^2 = (x_o - x_o')^2 + (y_o - y_o')^2 + (z_o - z_o')^2$.

What is the elapsed time between the two events? This question makes sense. Explicitly, the elapsed time would be given by the following equation: $(\text{elapsed time}) = t_o - t_o'$.

It is not shocking that these questions all make sense, for we are used to addressing them in everyday life. "We are now in the exact position where the *Titanic* sank." "It is now just six weeks since Carter was elected." It is in this sense, then, that the Aristotelian view is the popular one.

Finally, we can give a few additional examples of everyday notions which, within the Aristotelian view, make sense. "Is this particle at rest?" is a sensible question, for we can answer it by finding the position of the particle at various successive times. If that position is always the same, no matter what the time, we may say that the particle is at rest. "What distance did this particle travel between one time and some time later?" is sensible. At each instant of time, we consider the event "the particle at that instant," and associate with that event, by the Aristotelian view, a position in space. Computing the distance between these successive positions in space, and taking their sum, we obtain the total distance traveled by the particle. "What is the speed of this particle?" Since we attach meaning to the distance traveled by a particle, we also attach meaning to its speed, the number computed by dividing distance traveled by elapsed time.

Not a single one of all the notions above will make sense, in their present generality, in relativity theory.

Here and hereafter, we shall denote by M the set of all possible events in our universe: all those events that have occurred in the past, all those occurring now, and all that will occur in the future; those in this room, in our solar system, in other galaxies. This one enormous set M will be called space-time.

A point of M, then, represents an event. A region of M, on the other hand, represents some collection of events, for example, the collection "all events which occurred within this room between 10:30 and 11:30 on 8 January 1976." As an illustration, we will now describe in terms of space-time the idealization involved in the original description of an event. Let a firecracker explode, and consider the collection of all events internal to the explosion itself. They would correspond to some region, as shown in figure 1. If, instead, a smaller, faster-burning fire-

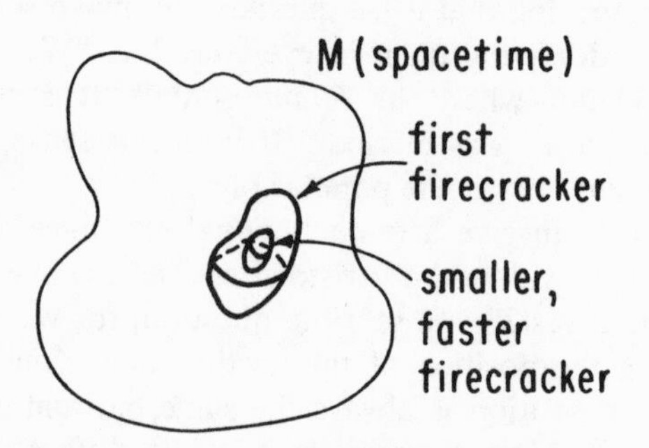

Fig. 1
The representation within space-time of the idealization implicit in the definition of an event. In the limit of "smaller, faster-burning" firecrackers, the region converges on a point of space-time.

cracker had been used, the corresponding region would be smaller in space-time. The idealization, then, involves the "collapse of these regions down to a single point of space-time."

Space-time would not be very interesting if it served merely as a repository for events. Its interest stems, rather, from the fact that many other, considerably more complicated things in the world can also be described within its framework. We shall see many examples of this later; we will give just one here. Let there be one particle, which we wish to describe in terms of space-time. Now a particle could not be described by a single point of M (that is, by a single event), for a particle has "extension in time." The appropriate description is in terms of a certain collection of events, namely the following. Consider the collection of all events which occur in the immediate presence of the particle (that is, for events marked by a firecracker, those for which the particle is internal to the explosion). This is the set of events which would be described if one continually followed the particle around throughout its life, snapping one's fingers on it. The resulting collection of events would be described by a line drawn in space-time (the line so drawn that it passes through precisely the events described above). This single line, called the world-line of the particle, completely describes everything one could want to know about the particle, for it tells us all the events experienced by the particle, that is, "where the particle is at all times." A particle, then, is not a "point" from the viewpoint of space-time—it is a line.

As an extension of this example, let us now consider two particles, A and B. Each particle is represented in terms of space-time by its world-line. Suppose that these two lines happened to intersect at some point p, as shown in figure 2. How is this to be interpreted physically? Well p, as a point of space-time, represents an event. The essential feature of p is that it lies on both world-lines. This means that the event represented by p is directly experienced by both particle A and particle B. In other words, both particles were there at that event. This is what we would call a physical collision of two particles. Intersection of world-lines thus corresponds to collision. If the world-lines of two particles do not intersect, the particles never collided.

We are trying to discover a theory of relationships between events. What role does space-time play in this endeavor? Since the events are represented as points of M, relationships between events are relationships between the various points of M. A set of such relationships is a kind of internal structure imposed on M. Our goal, then, is to find what structure we can on space-time M.

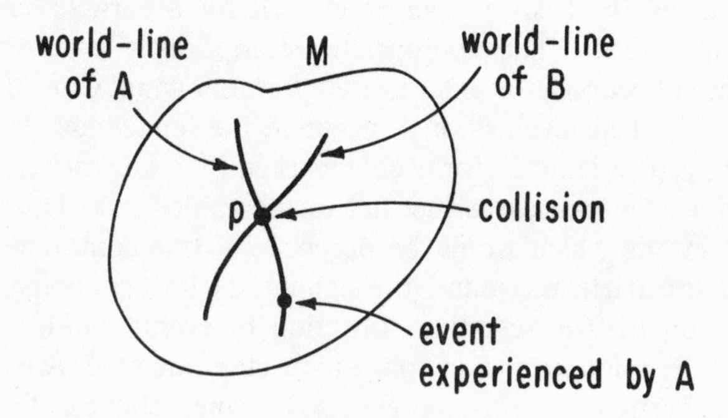

Fig. 2
The representation within space-time of the collision of two particles. The event "collision" is the intersection of their world-lines.

2

The Aristotelian View: A "Personalized" Framework

In chapter 1, we introduced, on the one hand, some elements of the Aristotelian view, and, on the other, space-time. The former is a particular attitude about what natural relationships there are between events, the latter is a more or less attitude-independent assemblage of events. Our first goal in the present chapter is to combine these two. That is, we wish to express the relationships between events according to the Aristotelian view as structure within space-time. We then give, using this "Aristotelianized space-time," several examples of how one translates back and forth between geometrical constructs in space-time and various goings-on in the physical world.

What is it that the Aristotelian view provides? It provides a characterization of each event by four real numbers, the values of x, y, z, and t. What is space-time? It is the collection of all possible events. Thus, the incorporation of the Aristotelian view into space-time is just the introduction, in space-time, of a certain coordinate system. (A coordinate system in the plane permits us to associate, with each point of the plane, two real numbers, the value of x and the value of y. Here, our coordinate system in space-time M permits us to associate, with each point of M [event], four real numbers, the values of x, y, z, and t.)

When we say "The plane is two-dimensional" we mean essentially that, to locate a point of the plane requires the specification of two real numbers (the values of x and y). Similarly for "Physical space is three-dimensional." By the same token, then, we are to regard space-time as four-dimensional, for the location of a point in space-time (event) now requires the specification of four numbers. One remark should be made in connection with this four-dimensionality. The view has for some reason come to be widely held that "the fourth dimension" is a deep and mysterious thing which permits extraordinary happenings in the world, and which only a few people can really understand. We emphasize that this is just not true. We now already "have four dimensions." On the other hand, we have not yet introduced a single statement about the way the physical world operates that was not known to all of us since childhood. True, we have perhaps been more careful and precise in our discussion than we might have been previously, yet the fact remains that, with no additional contributions whatever to our basic fund of physical information, we have arrived at a description in terms of four dimensions. If you like, "four dimensions" is just a convenient way of describing the world and thinking about the world, nothing more. Is the "fourth dimension" real? It should now be clear, from these remarks and from the discussion of "reality" in chapter 1, that physics will not answer such a question, and that the attitude of physicists will be that such a question is not germane. There is the physical world, and then there is our description of it. As long as our description is reasonably clear and reasonably accurate, there will be no objections. We can change our description every Friday morning if we wish. Nature doesn't care about our descriptions; She just keeps rolling along. If these days we choose to describe Her in terms of a space-time, and if that space-time has four dimensions, then, as long as that description is reasonably clear and reasonably accurate, that's fine and that's the end of it. Tomorrow's description may have two dimensions or nineteen dimensions. All of us, I can assure you, now understand "the fourth dimension" as well as anybody.

We now have our space-time M, with a coordinate system. In order to see relationships quickly, and in order to avoid

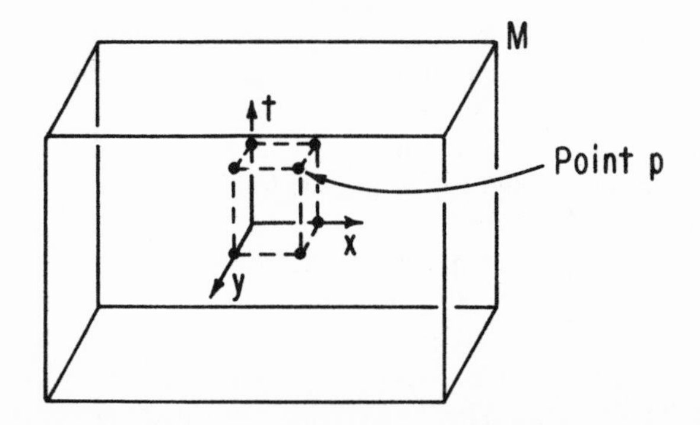

Fig. 3
The location of a point p in space-time by its Aristotelian coordinates.

lengthy and wordy discussions, it is convenient to draw pictures (the same reason, we remark, that pictures are drawn in plane geometry). We shall conventionally draw the picture shown in figure 3. Here, space-time is indicated by a "cube," and the coordinate axes are those shown. Why is the z-axis missing? The problem here is that we don't have enough room in our diagram to draw in all four axes, and so something had to be left out. We didn't want to leave out the t-axis, for it is essentially different from the other three, and we want to be able to represent interesting things associated with it. So, we just omit from the diagram one of the spatial axes. The result is that the diagram is not completely true to "Aristotelianized space-time," and, indeed, it can sometimes be a bit misleading. One quickly learns from experience, however, how to utilize what is valuable in such diagrams while avoiding what is misleading. We emphasize again the context of this problem: these pictures are merely a convenient way of looking at (not a necessary ingredient of) space-time, while space-time is merely a convenient way of looking at (not a necessary ingredient of) nature.

We wish next to understand the physical significance of certain lines and surfaces drawn in space-time. Consider first the t-axis. Now the t-axis, as a line in space-time, represents a certain collection of points (namely, those through which this line passes). That is to say, it represents a certain collection of events. What is the physical description of this collection? A point of space-time will be on the t-axis provided its x value is zero, its y value is zero, and its z value is zero. Recall, now, how an event is assigned, in the Aristotelian view, its four real numbers: The first three were those copied by the person experiencing the event from his badge, while the fourth was that read from his watch. An event will therefore have its x value, y value, and z value all zero provided the person experiencing that event wore a badge saying "$x = 0$, $y = 0$, $z = 0$." But this particular badge-marking just singles out a particular person in our assemblage of people filling the room. We conclude that the points on the t-axis represent precisely all events experienced by that person whose badge reads "$x = 0$, $y = 0$, $z = 0$." Now let us consider some other vertical straight line parallel to the t-axis. This line might consist, for example, of all points with x value 3, y value 7, and z value -2. Clearly, the points on this line represent precisely all events experienced by that person whose badge reads "$x = 3$, $y = 7$, $z = -2$." (Note here the translation from fig. 4, in which certain information must necessarily be left out, to the discussion, in which this information is restored.)

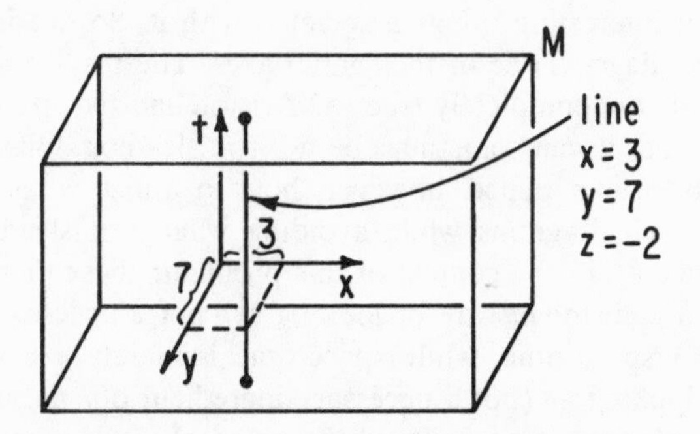

Fig. 4
The world-line of the individual whose badge reads $x = 3$, $y = 7$, $z = -2$ is the vertical straight line with these coordinates.

To summarize, each person who has helped fill the room has his own personal vertical straight line in the diagram, and each such line represents a person. To put matters another way, if we idealize each of our subjects as a "particle," then the vertical straight lines are the world-lines of these particles.

Consider next the plane given by "$t = 0$," that is, the plane consisting of all points with t value zero. Since within this plane we are free to specify the x value, y value, and z value arbitrarily, this "plane" is three-dimensional. (It takes the specification of three numbers to locate where one is on the plane.) To avoid confusion, we usually call such a surface a "3-plane" or "3-surface." We wish to interpret physically the events corresponding to points lying on this plane. Clearly, these are just the events with t value zero, that is, those such that the person directly experiencing the event did so just as his watch read "zero." We may describe the situation even more explicitly as follows. Suppose that the moderator had, say at time $t = -10$, made the following announcement: "I want each of you here in the room to keep an eye on his watch, and when your watch reads 'zero,' I want you to snap your fingers once." Then there would be silence at $t = -9$, at $t = -8$, and so on, until $t = 0$, at which time there would be one, very loud, "snap" (as everyone snapped their fingers), followed again by silence. Consider now the collection of all events that would be marked by all those finger-snappings. This collection would of course be precisely those events corresponding to points on the 3-plane $t = 0$. Consider now another horizontal 3-plane parallel to the 3-plane "$t = 0$," say the 3-plane given by $t = 11$ (fig. 5). Clearly, this 3-plane passes through points corresponding to those events which would be obtained by a similar announcement by the moderator, but his announcement would now be ". . . watch reads '11.' . . ." To summarize, the horizontal 3-planes represent "watch readings."

We may restate the above as follows: Each vertical straight line corresponds to one of our individual subjects, that is, it corresponds to what we have called a "position in space." Each horizontal 3-plane corresponds to a "watch reading," it corresponds to what we have called a "time of occurrence." Thus,

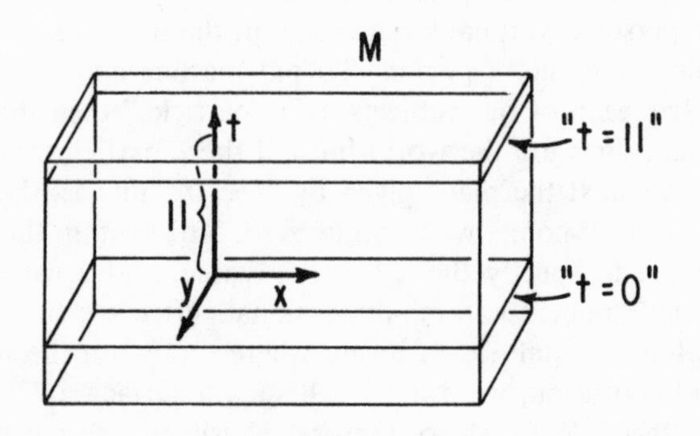

Fig. 5
The horizontal 3-plane given by $t = 11$ in space-time represents "all space at time $t = 11$."

"positions in space" are represented, no longer by "points in space," but rather by certain lines in space-time. "Times" are now represented by certain 3-planes in space-time. We may now go back and interpret geometrically our characterization of the Aristotelian view. According to this view, "an event is naturally characterized by giving its position in space together with the time of its occurrence." This sentence is now reformulated in terms of our space-time thus: "A point of space-time is naturally characterized by giving the vertical straight line on which it lies together with the horizontal 3-plane on which it lies." This, of course, is right, for the point so characterized would be that lying at the intersection of the line and plane.

In chapter 1 we discussed a number of relationships between events which are available in the Aristotelian view. We now wish to reformulate these various relationships geometrically within space-time. "Do two events have the same position in space?" translates to "Do two points of space-time lie on the same vertical straight line?" "Do two events occur at the same time?" translates to "Do two points of space-time lie on the same horizontal 3-plane?" The other two questions—"What is the spatial distance between two events?" and "What is the elapsed time?"—are slightly trickier. Let p and q be two events as shown in figure 6. Their "positions in space" are represented by the vertical straight lines on which they lie. To compute the spatial distance between the events, we must therefore compute the geometrical distance between the corresponding lines. Similarly, to compute the elapsed time, we compute the geometrical distance between the corresponding horizontal 3-planes. Note that neither of these numbers is the same as the geometrical distance between the actual points in the diagram.

Let us fix attention for a moment on a single particle in the world, and let us draw (fig. 7) its world-line in space-time (that is, we recall, that line passing through those points corresponding to events directly experienced by the particle). We are interested in the following question: How does one figure out what the particle is doing physically by examining only its world-line? We begin by drawing in one of the horizontal 3-planes, say that corresponding to time $t = 7$. Consider the

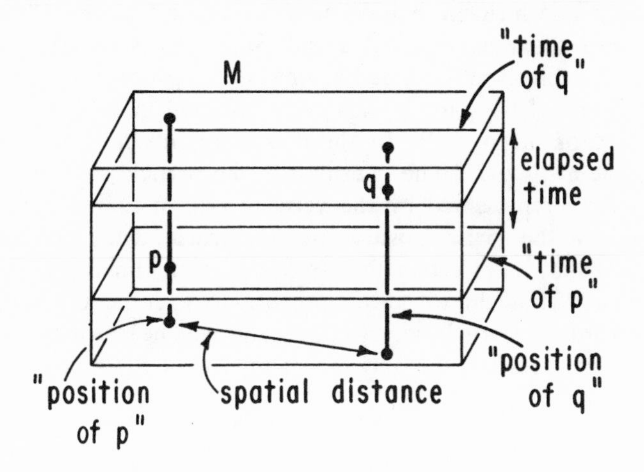

Fig. 6
Points p and q of space-time are described, in the Aristotelian view, by their positions in space (vertical lines) and their times of occurrence (horizontal 3-planes). The spatial distance and elapsed time between the events is then given, respectively, by the distance between the vertical lines and the distance between the horizontal 3-planes.

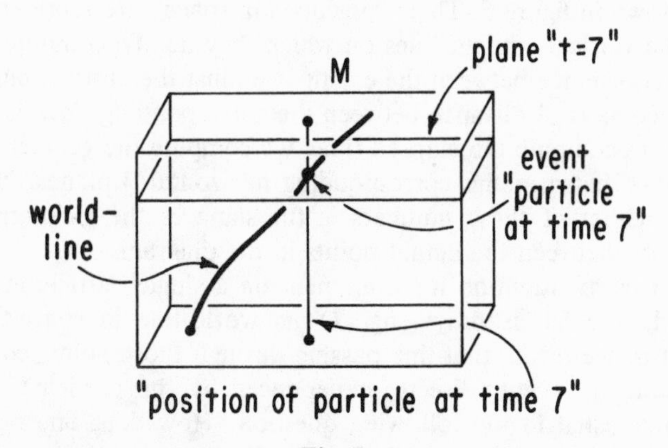

Fig. 7
The dynamics, according to the Aristotelian view, of a particle in space-time.

point p at which the world-line intersects this plane. What is the corresponding physical event? Since p lies on the plane "$t = 7$," this event occurred at time $t = 7$. Since p lies on the world-line of the particle, this event occurred in the immediate vicinity of the particle. In short, p represents the event "the particle at time 7." The "position of the particle at time 7" is of course represented by the vertical straight line through p. Similarly, the spatial position of the particle at other times is obtained by intersecting the world-line with other horizontal 3-planes. Thus, the world-line provides us with the information of where the particle was (and is, and will be) at every possible time, which is the only information we care to have about the particle.

Figure 8 shows the world-lines of four different particles, A, B, C, and D. The world-line of particle A is a vertical straight line. Hence, this particle is at rest, for the points which result from intersecting this world-line with various horizontal planes all lie on the same vertical straight line (namely, the world-line of the particle), whence the particle maintains always the same position through time. The world-line of particle B is straight but not vertical. This particle is therefore moving. Its positions

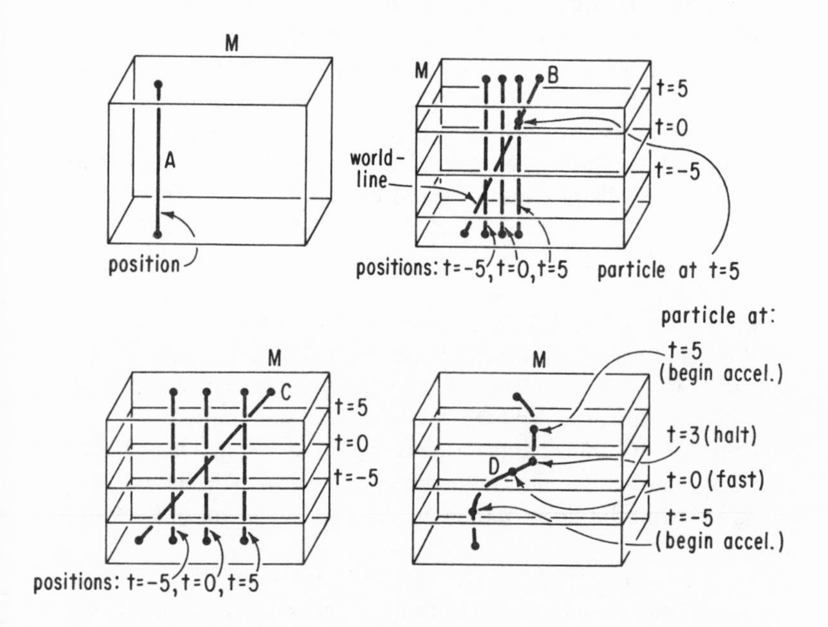

Fig. 8
Four examples of the interpretation of the space-time diagram of a particle. Particle *A* is at rest. Particle *B* is moving at a small, constant velocity, and particle *C* at a larger, constant velocity. Particle *D* is executing a more complicated motion.

at the three times—$t = -5$, $t = 0$, and $t = 5$—are represented by the vertical lines shown. We see that successive vertical lines are displaced from each other in the x-direction, thus the particle is moving in the x-direction. Since these lines (corresponding to positions after equal time increments) are equally spaced, the particle is moving with constant speed. The world-line of particle C is also a straight line, but it is "tipped more" than that of B. As a consequence, the spacings between positions at times $t = -5$, $t = 0$, and $t = 5$ (that is, between the three vertical lines) are greater. This means that particle C has more "change of position per amount of time" than B. Physically, we would say that particle C is going faster than B. The motion of particle D is even more complicated. Up until time $t = -5$, the particle was at rest. Then, beginning at time $t = -5$, it began accelerating in the x-direction. By time $t = 0$ it had gotten up to a rather large speed (as indicated by the fact that the segment of the world-line as it passes through the plane $t = 0$ is tipped far from the vertical). The particle then continued along at this large constant speed until time $t = 3$, when it suddenly came to a halt. The particle then remained at rest between $t = 3$ and $t = 5$. At $t = 5$, however, the particle suddenly decided that it wanted to accelerate in the other direction (toward decreasing x values), and it did so, moving in that direction. We see then that vertical straight world-lines correspond to particles at rest, nonvertical straight lines to particles moving at constant speed (where the speed is determined from the slope of the line), and accelerating particles to curved lines. Furthermore, the instantaneous speed of a particle is determined from the slope of the corresponding segment of its world-line.

As a second example, we analyze figure 9, in which there are two particles, A and B. Initially, A was at rest, and B was moving quickly toward A. This is seen, for example, by examining the positions of the particles at times $t = -5$ and $t = -3$. In these two seconds, A has kept its position while B has moved considerably closer. The event p, the intersection of the world-lines, represents the collision of the particles (they experience an event in common). The particle B "bounced off A," that is, it began to move (at a slower speed, incidentally) in the opposite direction. Particle A, on the other hand, "recoiled from the collision," that is, it began to move in the direction in which B was originally moving.

Aside from the technical details (which are also important), there is an important general observation to be made from the discussion above. It is this. There is no dynamics within space-time itself: nothing ever moves therein; nothing happens; nothing changes. In figure 9, for example, a certain amount of action is represented—particles moving about, collisions, and so on. Yet this dynamic, ongoing state of affairs is represented, past,

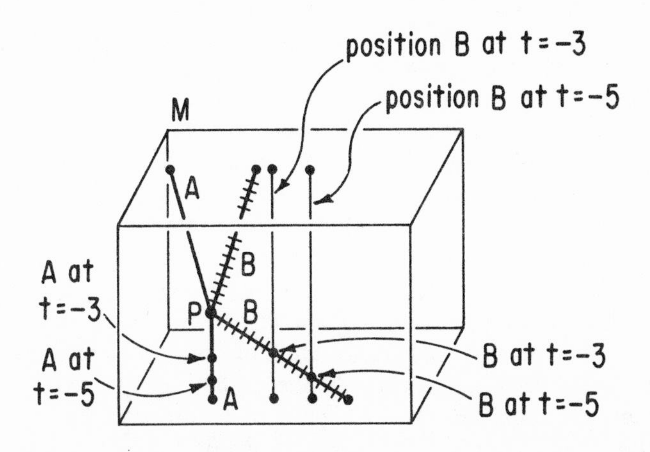

Fig. 9
The space-time diagram of the collision between two particles.

present, and future, by a single, unmoving space-time. Imagine a film has been taken of what occurs in the world, that this film has been cut into its individual frames, and that these have been stacked on top of each other. The result is similar to space-time. In the case of the film, "dynamics" is only recovered by comparing successive frames in the stack; in the case of space-time, by comparing the situation as recorded on several horizontal 3-planes. In particular, one does not think of particles as "moving through" space-time, or as "following along" their world-lines. Rather, particles are just "in" space-time, once and for all, and the world-line represents, all at once, the complete life history of the particle.

Had our world consisted of nothing but individual particles, our description of objects within that world would now be completed. But there are, fortunately, many other types of objects in the world. We now discuss how some of these are to be treated within the context of space-time.

Consider first a piece of rope or a stick (idealized to be very thin, just as events were idealized to have "extension in neither space nor time," and particles were idealized to "have no extension in space, while living through time"). We describe this rope by the collection of all events which occur in the immediate presence of the rope (that is, physically, we imagine a crowd of people standing around the rope during its life, continually snapping their fingers over all parts of the rope). The result is a certain collection of events, a certain set of points in space-time. The resulting set will be a certain two-dimensional surface in space-time, as shown in figure 10. Let us analyze what the rope corresponding to this figure is doing physically. The edge of the surface marked A represents the world-line of one end of the rope (where we think of this end of the rope as corresponding to a particle). Similarly, the edge marked B represents the other end of the rope. Let us now consider a particular time, say $t = 1$. The corresponding horizontal 3-plane is drawn in figure 10, and its intersection with the surface describing the rope is the line shown. Seen from above, this 3-plane and line would have the appearance shown in figure 11. What do the points on the line mean physically? Each such point represents some event. Since the point lies on the surface describing the rope, the corresponding event occurs in the im-

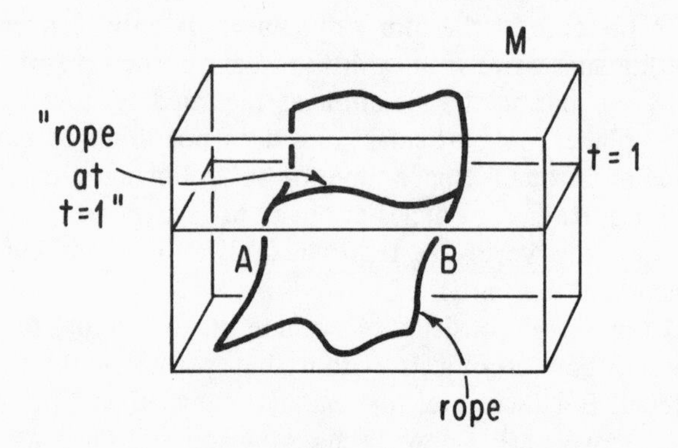

Fig. 10
The space-time diagram of a rope is a two-dimensional surface.

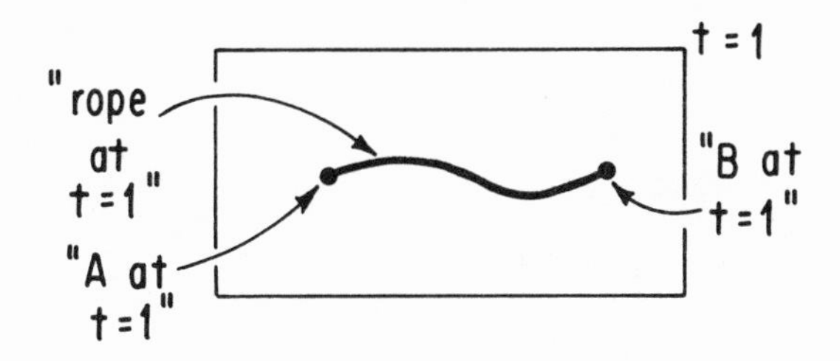

Fig. 11
To obtain the spatial configuration of the rope of figure 10 at a given time, one intersects the world-surface of the rope with the horizontal 3-plane representing that time. Here, then, is that spatial configuration at time $t = 1$.

mediate presence of the rope. Since the point lies on the horizontal plane "$t = 1$," the event occurs at time $t = 1$. In short, the points on this line represent the events "on the rope at time 1." The line above, then, represents "the spatial configuration of the rope at time 1." Similarly, introducing other horizontal 3-planes, and intersecting these with the surface describing the rope, we recover "the spatial configuration of the rope at all other times." In short, we can determine from its surface in space-time how the rope is shaped and where it is at all times, which is all we care to know about the idealized rope. This two-dimensional surface, which describes completely what the rope is doing, is called the world-surface of the rope.

Some examples of world-surfaces of ropes are shown in figure 12. The rope illustrated in figure 12,*A* begins at rest, straight along the x-direction. At time $t = 0$, the rope begins to move in the x-direction, that is, along the direction in which it is pointing. The rope moves thereafter at constant speed. The rope in figure 12,*B* is also straight in the x-direction (for any horizontal 3-plane, intersected with the world-surface of the rope, results in a straight line-segment along the x-direction). Now, however, the rope is moving in the y-direction, that is, in a direction perpendicular to itself (for the straight line-segments which result from intersecting the world-surface of the rope

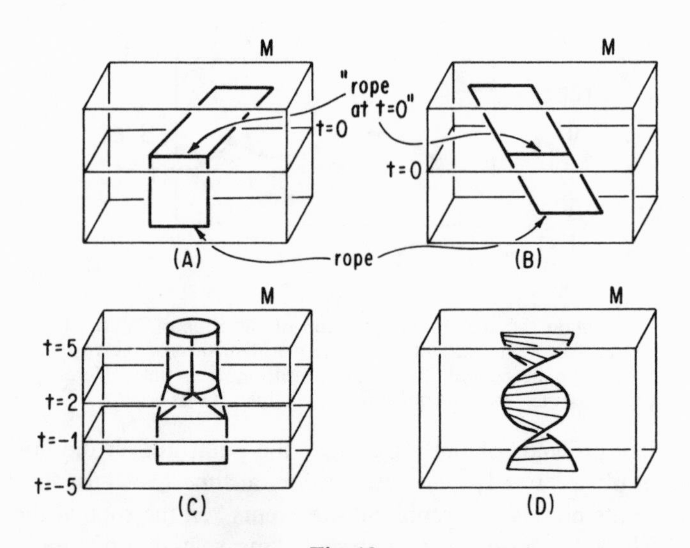

Fig. 12
Four examples of the interpretation, in the Aristotelian view, of the world-surface of a rope. The rope in (A) is initially at rest, but then begins to move in the direction of its extension. The rope in (B) moves at constant velocity orthogonally to the direction of its extension. The rope in (C) forms itself into a circle. Finally, the rope in (D) is rotating about its midpoint.

with successive horizontal planes are displaced from each other in the y-direction). The rope is moving with constant speed. The rope in figure 12,C was at rest, stretched out in the x-direction, until time $t = -1$. At that time, the rope began to curve around, in the spatial x-y plane, to form a circle. The formation of the circle was finally completed by time $t = 2$, and the rope thereafter remained at rest in this circular configuration. The rope in figure 12,D always remains straight, but is rotating about its center, at constant speed, in the spatial x-y plane.

An object in one higher dimension, such as a sheet of paper, is characterized, similarly, by the collection of all events which occur in the immediate presence of the paper. In this case, the corresponding set of points in space-time would lie on a three-dimensional surface. The location and configuration of the paper at any instant of time would be recovered from the world-surface of the paper by intersecting that surface with the horizontal 3-plane corresponding to that instant of time. In this way one recovers from the surface where the paper is and how it is shaped at every instant of time. It is, unfortunately, difficult to draw reliable pictures in this case, because of the dimensional limitations for the picture we draw of space-time.

We come finally to solid objects (objects which "occupy space," which "have volume") such as the earth. Again, such objects are described by the collection of all events which occur in their immediate presence (that is, in the case of the earth, by all events which occur on the surface or in the interior of the earth). The corresponding set of points in space-time is called the world-region of the object. Once again, we determine the location and shape of the object at any instant of time by intersecting its world-region with the horizontal 3-plane corresponding to that instant of time. Finally, again, the information of the location and shape of such an object at every instant, obtained as above, tells us all that we presently wish to know.

Figure 13 gives some examples of world-regions of planets. The planet in figure 13,*A* is spherical and at rest until time $t = 0$, at which time it begins moving, at constant speed, in the x-direction. (The intersections of the world-region of the planet are solid disks—one dimension suppressed, remember.) These disks are all identical until $t = 0$, after which successive disks are displaced in the x-direction, indicating motion of the planet. The planet in figure 13,*B* was at rest and spherical until time $t = -1$, at which time it began changing to an ellipsoidal shape. This change in shape was finally completed by time $t = 2$, and thereafter the planet remained in this shape (with the longer axis of the ellipsoid pointing along the x-direction). The planet in figure 13,*C* is oscillating—expanding and then contracting and then expanding, and so on. It always retains its spherical shape. Finally, the planet in figure 13,*D* remained spherical and at rest until time $t = -1$. At that time, the planet exploded

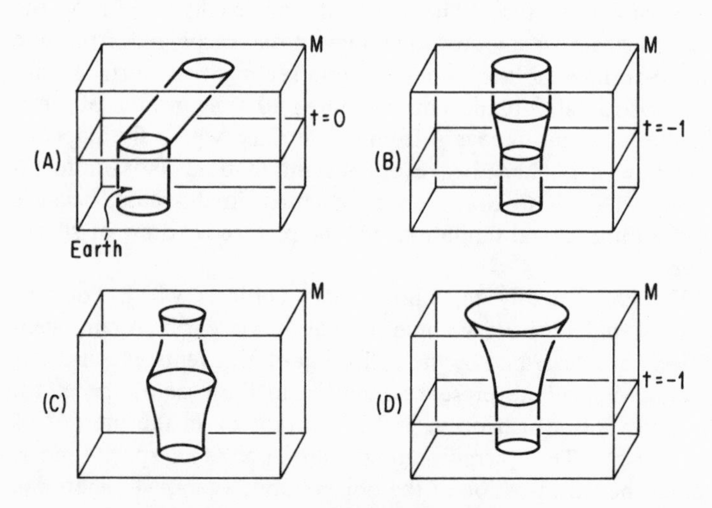

Fig. 13
Four examples of the interpretation, in the Aristotelian view, of the world-region of a planet.

—it began expanding more and more. The explosion was spherically symmetric. (In this case, for example, the intersections of the world-region with successive horizontal 3-planes are solid disks, all at the same location, until time $t = -1$. Thereafter, successive disks are larger, indicating that "the spatial size of the planet became larger and larger through time.")

Of course, it is not necessary to have only one of the objects above in space-time—one can have several simultaneously. Figure 14 shows some typical examples. In figure 14,*A*, we see the world-region of the earth, and the world-line of Chicago (idealized, here, as a "particle"). Of course, were Chicago underground, its world-line would be within—rather than on—the cylinder. This figure also shows the world-line of a traveler who lives in Chicago and takes a trip around the earth, leaving at time $t = -1$ and returning at $t = 4$. The world-line of a satellite orbiting the earth is also shown. In figure 14,*B* we have the world-region of the earth. Shown also is the world-line of a person who stays at one place on the earth, and the world-line of a bird, initially in a tree above the earth. The person throws a rock which strikes the bird, knocking it out of the tree. The events p, q, r, and s represent, respectively, "throwing of the rock," "rock striking bird," "rock meeting ground," and "bird meeting ground." Note that, according to this figure, the rock falls quickly to the ground, while the bird flutters down more slowly, for event s is later than r. Figure 14,*C* shows a bat (idealized as a straight stick) being swung and striking a ball (idealized as a particle). Note the event "impact of the bat and ball." We will postpone for a few moments discussion of figure 14,*D*.

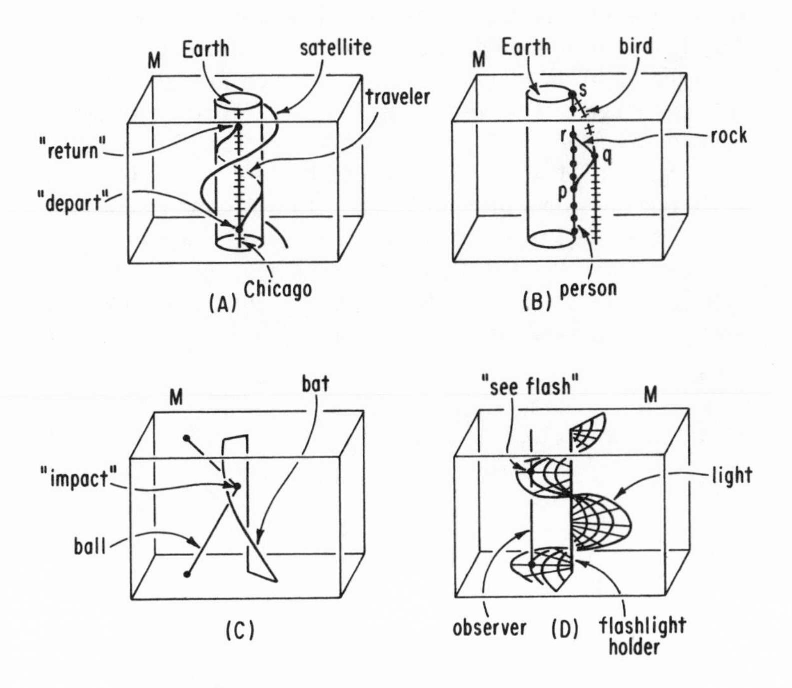

Fig. 14
Four examples of the interpretation, in the Aristotelian view, of more complicated space-time diagrams.

This long discussion and these many examples are intended to serve two purposes. On the one hand, we are trying to make a case that space-time is an appropriate carrier of all information about what is going on (has gone on, will go on) in the world, that it has enough structure, enough richness for the job. On the other hand, the examples are intended to give some experience in the translation from geometrical constructions in space-time to physics. It is important that one acquire some facility with this technique, that one be able to make these translations quickly and effortlessly.

Finally, we discuss here one further topic, a topic which will play an important role later, light. Let us first consider the situation in which someone carries with him a highly directional flashlight (that is, the light-beam does not spread out). Let him keep the flashlight off, then turn it on for an instant, and then turn it off again. The result will be a small "particle of light" emitted from the flashlight. We would represent this situation as in figure 15,A. Note the event "flashlight flashed." In figure 15,B, our individual turns the flashlight on (event p), leaves it on for a while keeping it always pointing in the x-direction, and then turns the flashlight off (event q). The 2-surface in that figure represents all events which are illuminated by light from the flashlight (just as the tilted line in fig. 15,A, the

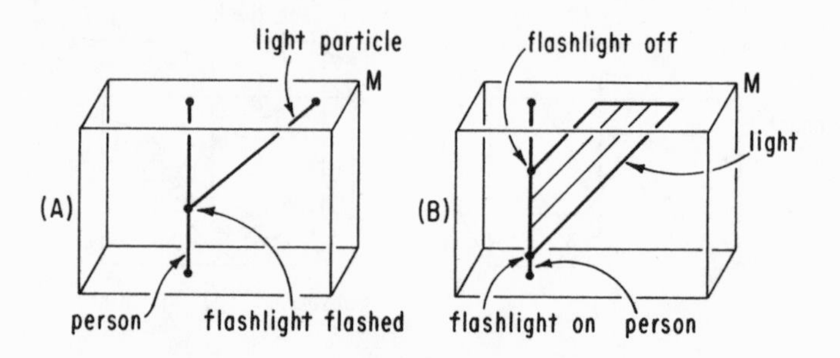

Fig. 15
The space-time diagrams representing (A) the emission of a momentary flash of light in a given direction, and (B) the emission of a beam of light in a given direction.

"world-line of the particle of light," represents all events there illuminated). Light, then, is described by the collection of all events illuminated. We may now return to the interpretation of figure 14,*D*. The vertical line at the center represents the world-line of a person (at rest) holding a flashlight. The flashlight is always on, and the person holding it is turning around in the *x-y* spatial plane, acting like a searchlight. The two-dimensional surface is the collection of all events illuminated by the light. The other vertical line is the world-line of some other person (also at rest). Note the events at which this line meets the surface representing the light. These, of course, are the events at which our second person sees the searchlight sweep past him.

We now consider, finally, a slightly different setup involving light. Let our subject now carry with him a light-source which sends out light in all directions (for example, a light bulb). Let him keep the source off, then turn it on for one instant, then keep it off thereafter. Again, we can represent the emitted light by the collection of all events illuminated. What will the corresponding surface in space-time look like? There are at least two ways of arriving at the answer. Let *p* be the event "bulb flashes." On the other hand, we could regard the light emitted by the flash as a bunch of particles of light, going in all spatial directions from *p*. Each of these "particles of light" would be described by its world-line, a tilted straight line beginning at *p*. The corresponding surface, the collection of all events illuminated, would then be just the union of all these lines. The result would be the conical surface shown in figure 16. Alternatively, one could have arrived at such a conical surface as follows. We intersect this surface with successive horizontal 3-planes. The plane slightly above that through *p* intersects our surface in a small circle, slightly higher planes ("later") in slightly larger circles, and so on. These circles (spheres physically, one dimension suppressed) represent "all events illuminated at one instant of time." Thus, the cone represents physically a "spherical surface of light, which is expanding outward with time." But this is precisely what would happen physically were our light bulb flashed. The flash of light would be a spherical surface, expand-

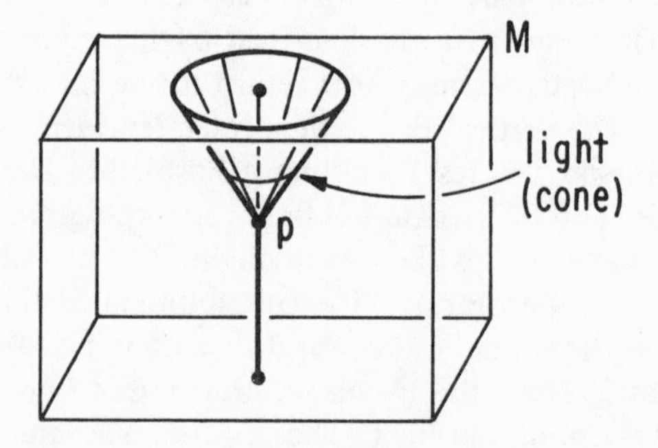

Fig. 16
The light-cone of event p is the locus of all events illuminated by a flash of light emitted in all directions from p.

ing outward with time from the bulb, which would be always at the center of the successive spheres.

This conical surface, which describes the behavior of a flash of light emitted in all spatial directions at event p, is called the light-cone of p.

The cube that we draw in representing space-time is not intended to represent a "boundary" of space-time in any sense—space-time is supposed to extend beyond that cube in all directions. It serves merely the artistic role of making it easier to show geometrical relationships between points, lines, and so on in space-time. We shall often leave out the "cube" when it is not necessary for clarity.

We now conclude our discussion of the Aristotelian view. We will begin with a few general remarks about this view and what it means, and end with a statement of why this view is felt to be inappropriate.

The first point to be made is that a "view" (of the structure of space-time, such as the Aristotelian view) is a rather different kind of thing from a "theory of physics." The latter is supposed to give a fair amount of detail about the way certain objects or systems in the physical world operate, whereas the former is so broad and all-encompassing that it admits almost arbitrary "modes of operation." In short, the Aristotelian view is merely a device which permits a description of what is happening (has happened, will happen), without itself placing any particular restrictions on what happens. A theory of physics must take over from here to tell us what does (should?) happen. If the interest of the theoretical physicist is theories of physics, why do we bother with "views"? The point is that the structure one places on space-time is to provide a broad framework within which one's theories are to be expressed, thought about, and tested.

Nonetheless, the structure one decides is appropriate on space-time does have an influence—indeed, a strong influence—on which theories one considers. For one thing, one's theory must "make sense" within one's broader view of the structure of space-time. By this we mean that one's references to space and time (that is, to relationships between events) within the statement of the theory must be only to those relationships which are actually available according to one's view of the structure of space-time. (Thus, a theory which makes reference to the "spatial distance between events" will make no sense in the context of a space-time view in which "spatial distance between events" is not meaningful.) As examples, we here discuss two theories of physics (perhaps more properly, fragments of theories) in relation to the Aristotelian view. We emphasize that we are not here concerned with whether or not these theories are "true" (in the sense that their predictions are accurately reflected in the world), but rather with whether or not they "make sense" in the Aristotelian view.

1. Newtonian law of gravitation. At any one instant of time, the gravitational force between any two bodies in the universe is proportional to the product of their masses and inversely proportional to the square of the distance between them at that time. This law makes two references to space-time. Consider two bodies, described by their world-lines. First, we must be able to introduce events which correspond to "the locations of those bodies at one instant of time." Second, we must be able to compute a number which represents "the distance between

the bodies at that time." We now claim that the Aristotelian view permits both. Indeed, the two events which correspond to "the locations of those bodies at one instant of time" are obtained by drawing the horizontal 3-plane representing that "instant of time," and finding the points of intersection of that plane with the world-lines. "The distance between the bodies at that time" makes sense, since we can in the Aristotelian view associate with any two events a distance between them. Newton's law of gravitation, then, makes sense in the Aristotelian view.

2. Law of light. A small pulse of light travels at a speed of 3×10^{10} cm/sec. We describe the small pulse of light by its world-line. Does its "speed" make sense? The answer is yes, for we have seen earlier in this chapter that one can compute, given the world-line of a particle, a corresponding speed for that particle. The law of light makes sense in the Aristotelian view. (Although one could give examples of "laws" which do not make sense in the present view, all of these seem to sound rather contrived, perhaps another indication that the Aristotelian view is that of everyday experience.)

Our second point concerns the relationship between the Aristotelian view and space-time itself. It is intended that certain of the statements we have made so far are to refer just to the general idea of space-time, without reference to a particular "view," while others refer explicitly to the Aristotelian view. Examples of the former include, "A particle is described by its world-line." "A rope is described by its world-surface." "The intersection of world-lines corresponds to collision of particles." These are broad, qualitative statements about the physical interpretation of certain geometrical objects in space-time. Examples of the latter include, "To obtain the shape of the rope at an instant of time, intersect its world-surface with the appropriate horizontal 3-plane." "These two events occur at the same time." "The particle represented by this world-line is at rest, that by that world-line is moving with constant speed." These statements all require an ability to characterize events by their location in space together with the time of their occurrence. We have not always pointed out explicitly in which context our statements lie, and so it is a useful exercise to go back and make this decision for various of our statements. We could, of course, have proceeded more logically, first introducing space-time and its general features and then the Aristotelian view and its more specific features. Indeed, were we to start over at this point, knowing what we know now, we would do precisely this. We did not do this because it would have made our discussion of space-time seem rather abstract and nebulous—it would have been hard to see what we were driving at.

The third point is historical. It is not our intention to proceed historically, with the consequence that we now have a mix of very old and rather new ideas. The notion that occurrences are characterized by a position in space together with a time of occurrence is presumably a very old one. Essentially everything else is more modern: isolating the notion of an event, considering space-time, expressing goings-on in the physical world in terms of collections of events.

The fourth point concerns revolutions. We remarked in the introduction that the space-time point of view can be regarded as a revolution in our conception of the physical world. Having come this far, we can now state in a bit more detail what the revolution was and how far we have come along it. I wish to regard this revolution as having three components: (1) the decision to introduce events and space-time and to describe the world within this framework; (2) the decision that the question of what is the appropriate structure on space-time is an active one, and that one should worry about what the answer is to be; (3) the decision that the appropriate structure is that prescribed by general relativity. We have now essentially treated the first component. We have also at least entered into the second component. While nobody has lost any sleep yet, our worry will very shortly begin to intensify. We have not yet even touched on the third component.

The Aristotelian view seems simple and comfortable. Why do we not simply go along with it, and stop here? What is wrong with this view? Consider, as an example, the following situation. You are placed alone far out in space—far from our planet, from the sun, from other stars, from the galaxy—and are equipped with a small rocket that you can use to maneuver about. A stranger passes by carrying two firecrackers, which he explodes, one five seconds after the other. Having very little else to do as the stranger recedes into the blackness, you set for yourself the problem of deciding whether or not these two events occurred "at the same position in space." There is a simple resolution of this problem. You just locate those two events with respect to yourself, and regard them as having the same position if they had the same position with respect to you. This is precisely what the Aristotelian view would require. The person would set up his own personal "x, y, z, t-labeling" of events in space-time, and would regard the two events as having the same position in space provided they had the same x values, the same y values, and the same z values. In so doing, one should, however, recognize on what this decision depends. Suppose that, a week before the stranger's arrival, you had fired your rocket for a short while, giving yourself some speed in some direction. Had you done this, your decision as to whether the two events occurred at the same position might have been quite different (for, with this additional speed, you would have moved during the five seconds between the explosions). Alternatively, a second person out there viewing the explosions, this person moving with respect to you, might have a quite different opinion as to whether or not the two events occurred "at the same position in space."

In response to the discussion above, one could perfectly well take the attitude that that's the way it goes. "Occurred at the same position in space" could be regarded as a personal rather than a public issue, an issue which, unfortunately if you wish, will depend on who is answering the question, whether or not I fired my rocket a week ago, and so on. The issue, then, is that certain relationships between events admitted in the Aristotelian view depend on who is doing the observing and what that person has done in the past. The resolution offered is that we just learn to live with "many, personal Aristotelian views."

For the reasons discussed below, we do not wish to adopt this resolution. Physics as a whole can be viewed as a kind of struggle between what is to be internalized and what externalized. The ultimate issue, of course, is always what people will actually see or otherwise experience. If, however, we choose always to deal directly with this "ultimate issue," a great deal of confusion will set in. For example, we might have to consider several types of events, those seen by people who forgot their glasses (and to whom, therefore, the occurrence appears fuzzy) and those seen by people who brought theirs, those marked by firecrackers but heard by those hard of hearing, and so on. These extreme examples are intended to make the point that one will have difficulty getting anywhere in physics without at least facing up to the issue of some sort of separation of the actual, raw human experiences into some portion attributed to the person and some portion attributed to "Nature herself." So some separation of some type must be made to do anything. Physics, then, proceeds to make such separations. The lines are initially not always clear-cut, and their location occasionally gives rise to debate. As time goes on, however, the line becomes more or less tightly drawn, and the subject proceeds. One might well ask whether it will always be possible to pursue physics fruitfully by the drawing of such lines, or whether we may someday be forced to cease this activity. Perhaps we are even today, in quantum mechanics, being pushed toward the latter.

In any case, we do not yet seem to have these deeper problems (about whether such lines should even be drawn) in the case of the structure of space-time, but only the rather more straightforward problem of deciding where the line is to be drawn. The claim is that the line location suggested by the attitude above ("personalized Aristotelian views") will be inconvenient, for it robs space-time of essentially all "universal structure." If everything (the whole Aristotelian setup) is to be attributed only to individuals who do the observing, then nothing is left which can be regarded as "pure structure on space-time itself, without explicit reference to observers." One would much prefer to redraw these lines in such a way that space-time retains at least some—preferably, as much as possible—"universal, observer-independent structure." Put in another way, the Aristotelian view includes in part some universal structure of space-time itself and in part some internal features about the observer. What we wish to do is separate out the former and clearly post it within space-time. What is to happen to the latter? We shall describe observers, as we always have, by their world-lines in space-time. From the "universal structure" of space-time, together with the information of the world-line of the observer, we shall be able to recover the various actual experiences of our observer. The observer, then, will be reduced from "an Aristotelian setup" to "a world-line," with the additional information thus lost in the reduction permanently implanted in space-time.

3

The Galilean View: A Democratic Framework

We now wish to actually carry out the program outlined near the end of the previous chapter. The result will be an alternative view of space-time, which is called the Galilean view.

Our starting point will be a certain comparison of the characterization of space-time by one person with that by another. If, however, any such comparison at all is to be made, then we must first adopt some basis for comparison. That is, we must agree on some notions which are to be taken as "universal," common to all. It will then be through such universal notions that we can translate from one characterization to another. As the central universal notion, we choose that of an event itself. That is to say, we shall regard the marking of an event as something which can be seen by all. All are to agree "that was an event being marked." Having made this convention, space-time, the assemblage of events, can also be taken as universal. The point then is that events and space-time will be taken to be "out there in the world"—people will not have their own "personal sets of events" or their "personal space-time." Events and space-time will be the common ground on which various observers will meet.

We first consider the situation entirely from the viewpoint of a single observer (say, "us"). We suppose, then, that we have constructed our usual Aristotelian setup. We have distributed badges and watches, we label events by a "position in space" and a "time of occurrence," we draw our picture of space-time.

Within our picture we may, as we have seen, describe the various types of phenomena which take place in the world. Let us now consider, and then so describe, the following, particular, phenomenon. Let some other person organize his own group of people, distributing to them badges and watches as we have discussed earlier. Let this second group, however, be moving past us (we really mean *through* us) with constant velocity. This then is a particular "phenomenon in the world" (a bit more complicated than, but not essentially different from, the phenomenon "a particle"). Our problem is to describe it within our picture.

Consider first a single one of the individuals in the other group moving by. This particular individual (idealized, as usual, to a "particle") is completely characterized by a certain collection of events, namely those which occur in his immediate presence. To describe this individual in our diagram, therefore, we would simply draw his world-line (fig. 17). Since he is moving at constant velocity, this world-line will be straight but not vertical, as shown. Similarly, another individual in the other group would be represented by another line in our diagram. Since this second individual is moving with the same constant velocity, his world-line would be straight, and parallel to the first. Drawing in, then, the world-lines corresponding to all the individuals in the other group, we obtain a family of parallel straight lines which completely fills our diagram of space-time. This, then, is how we would represent geometrically the individuals in the other group.

We consider next how we would represent the watches they carry. Let us suppose that the moderator of the other group announces to his followers that they are to all snap their fingers at time $t = 0$. (To avoid unnecessary complication, let us also suppose that this group's watches are synchronized with ours.) As a result of these instructions by the other moderator, some fingers will be snapped, and some events will be marked. We would represent this collection of events within our diagram by drawing the line or surface or whatever which passes through the corresponding points in our diagram. In this case, since the

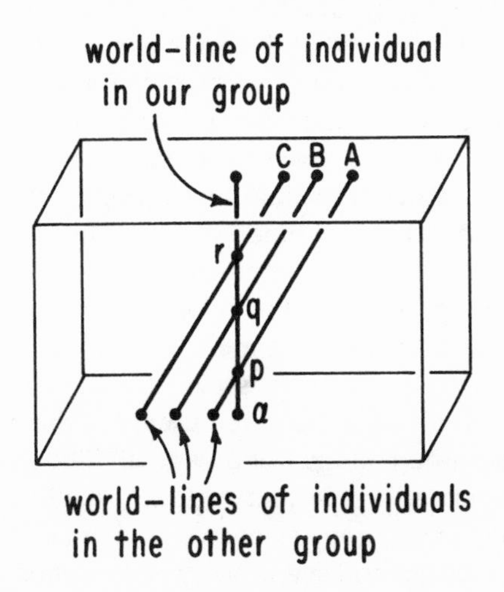

Fig. 17
The description, within our Aristotelian view, of a second Aristotelian setup. The world-lines of the members of the other group are straight and parallel, but are not, in general, vertical.

watches are all synchronized, the appropriate set of events in our diagram will be precisely the horizontal 3-plane that we would label "$t = 0$." Similarly, instructions by the moderator of the other group corresponding to a different t value would result in our drawing in some other horizontal 3-plane in our diagram. It is by means of these 3-planes, then, that we represent the watches of the other group.

So far, we have regarded this group of people, with their badges and watches, as just some phenomenon in the physical world, to be represented within our space-time diagram as any other phenomenon would be. We might now note, however, that this other group bears a remarkable similarity to us; they carry badges and watches just as we do. Indeed, they may, for all we know, be thinking of the same things that we are; they may actually regard themselves as an Aristotelian setup. What would they be saying, for example, about the collection of all events directly experienced by one of their people? They would say "Those events all occurred at the same position in space." (We, of course, would not share this opinion, for we represent this collection of events by a nonvertical line.) Similarly, the events marked when they all snapped their fingers at time $t = 0$ (according to their watches) would, according to them, represent "all space at the instant of time $t = 0$." (On this, at least, we would agree.)

We may now go even further along these lines, and try to reconstruct what their space-time diagram would be like. It would, of course, just be the usual Aristotelian's space-time diagram, with the world-lines representing the various individuals in their group straight and vertical. The surfaces representing "all space at one instant of time" would be horizontal 3-planes. This other group, of course, feels that it is at rest. It represents various phenomena in the world geometrically within its diagram.

Among the "phenomena in the world" seen by the other group is, of course, us. They feel that we are "moving by." Thus, they would be representing us geometrically in their diagram just as we have been representing them, that is, according to figure 18. The world-line α, for example, represents one of our people. Why are "the lines in their diagram representing our people" tipped the other way from "the lines in our diagram representing their people"? There are at least two ways of seeing this. On the one hand, it can be understood physically. It is just the statement that "If we, feeling that we are at rest, see the other group as moving in one direction, then they, feeling that they are at rest, see our group moving in the other direction." On the other hand, it can also be understood geometrically, and it is instructive to see how the argument works. Let us fix attention on three of the people in the other group, labeling them A, B, and C. The world-lines of these people, in our diagram, are those shown in figure 17. In their diagram (fig. 18), on the other hand, those world-lines are vertical. Let us now also consider one of the people, α, in our group. As we see in figure 17 (ours), the world-line of α meets the world-

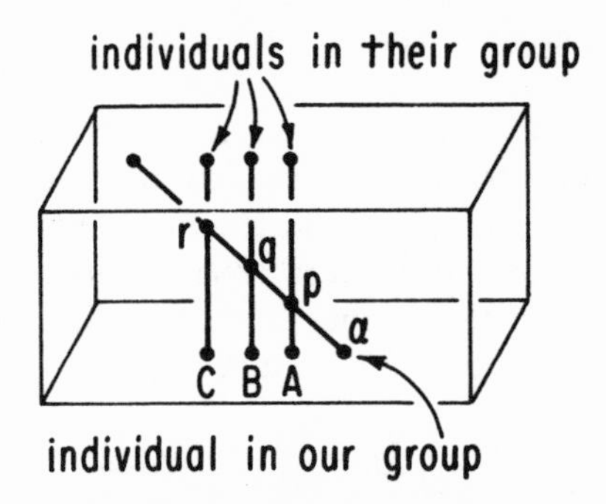

Fig. 18
Figure 17 as drawn by the other Aristotelian group.

lines of A, B, and C at three points, labeled p, q, and r, respectively. What this means in physical terms is that a certain individual, who calls himself α, sees himself collide with (better yet, *pass through*) first person A, then person B, and then person C. Now, the last statement above is simply a statement about world-lines and events—it is, if you like, an objective statement about the world, it does not refer particularly to a "view." This being the case, everybody must agree that person α met successively person A, person B, and person C. In particular, this same physical thing (the meeting of α with A, B, and C) must be shown in the space-time diagram of the other group (fig. 18). But we can now use this fact to decide how person α is to be represented in this last diagram. Whatever his world-line is to be, it had better meet first A, then B, and then C, that is, it had better pass through the points p, q, and r in that diagram. The particular line shown, therefore, must be the correct representation of the world-line of person α.

To summarize, we have one, objective "world out there," with its events. Two groups of Aristotelians look at this world, and each describes it in terms of a space-time diagram. Each group can, within its own diagram, describe any phenomenon taking place in the world. In particular, each group can describe itself (vertical straight lines, horizontal 3-planes) and also the other group (parallel nonvertical straight lines, horizontal 3-planes).

Clearly, these two diagrams are related to each other by a simple geometrical transformation. To obtain our diagram from theirs, for example, we just "slide the horizontal 3-planes over each other until our world-lines are straightened to the vertical." The result of this sliding is to tip their world-lines (fig. 19). Similarly, they recover their diagram from ours by "tipping

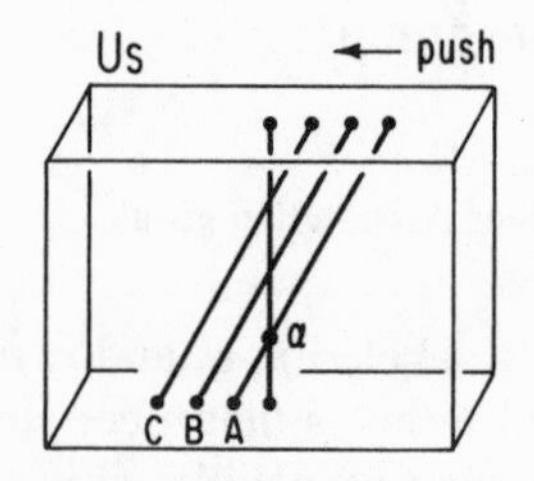

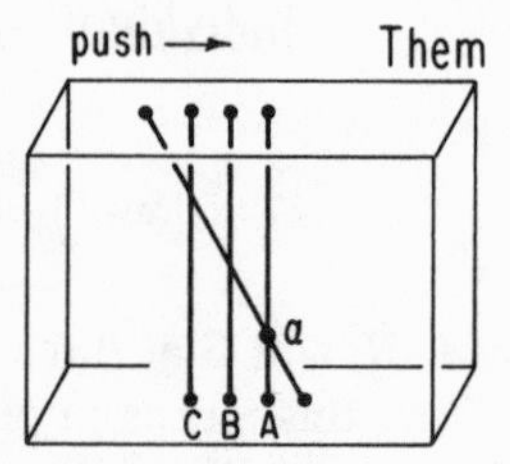

Fig. 19
The geometrical relationship between figures 17 and 18. The two figures are related to each other by sliding horizontal 3-planes over each other.

the other way." (The situation is similar to that of a single deck of cards, with several wires stuck through it from the top. By "beveling the deck" one can cause one or another of the wires to be vertical, tipping the others.)

So far, the only physical phenomena our two groups of observers have tried to describe have been themselves and each other. We now wish to consider other phenomena. In more detail, the situation we wish to consider is this. Something or other occurs in the world, and each of our two groups represents it within its own space-time diagram. We wish to find out how these two representations are related to each other. We emphasize that we are here concerned with a *single phenomenon* in the world, and the two different groups' description of it—the separate groups are not now setting up and studying their own private phenomena. It turns out that there are two (ultimately equivalent) ways of dealing with questions of this kind: a physical way and a geometrical way.

The physical way is the following. Let there be given, say in our diagram, various points, lines, surfaces, and so on which represent some arrangement of particles, strings, and so forth in the world. Since we know how to translate from the diagrams to the world, we may, if we wish, mark somehow the corresponding events in the world. But the other group also knows how to translate from the world to their diagram. They carry out this translation, and thus obtain various points, lines, surfaces, and so on in their own diagram. The translation process here, then, is from our diagram to the physical world to their diagram. (Note, here, the "universal" events providing the link between the two diagrams.) Of course we do not, in using this "physical way," have to actually physically mark the events in the world; rather, we make the transitions from our diagram to the world to their diagram in our mind's eye.

The geometrical way is rather easier. Consider some geometrical constructs (lines, surfaces, and so on) in our space-time diagram which are to represent some phenomena in the physical world. To obtain the geometrical constructs representing these same phenomena according to some other Aristotelian setup, we simply perform the "beveling the deck" operation illustrated above. That is to say, we imagine that the lines, surfaces, and so on have been marked on the cards in the deck, we "bevel the deck," and then we look at those marks again to see what the new lines, surfaces, and so on will be.

Since the operation of beveling is just a mapping from our description of actual physical events to their description, and since we describe everything in terms of collections of events, it is clear that these two methods are equivalent.

We will give an example of the use of these two methods. Consider figure 20,*A* (drawn, of course, according to our Aristotelian setup), showing the world-line of a particle, and, for reference, the world-lines corresponding to the other Aristotelian setup. What would the diagram for the second group be? Using the physical method, we would proceed as follows. What

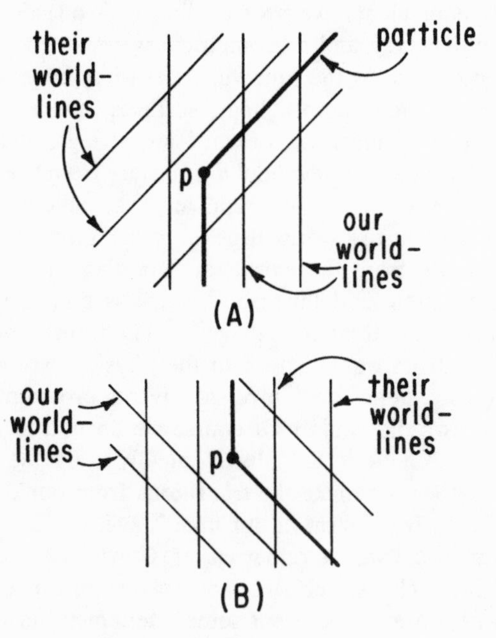

Fig. 20
Space-time diagrams for a particle which initially remains with a member of our group but, at event p, joins their group. Figure (A) is as would be drawn by our Aristotelian setup, (B) as by theirs.

does the diagram mean? It represents a particle which is at rest until some time (corresponding to the horizontal 3-plane through the event p), at which time the particle begins moving along with the other Aristotelian group. What words would the other group attach to this phenomenon? They would say "The particle was moving along with your group until some time (corresponding to the horizontal 3-plane through the event p), at which time the particle was stopped, remaining thereafter at rest." How would this physical description be translated into a space-time diagram? It would be figure 20,B (in which, for reference, the world-lines corresponding to our Aristotelian setup are drawn). This last, then, is the new diagram. In the geometrical method, one just takes the world-line illustrated in figure 20*A* and subjects it to the beveling operation illustrated in figure 19. The result, of course, is precisely figure 20*B*.

Two further examples of transitions from one diagram to another are in figure 21. Figures 21,*A* and 21,*B* show a collision between two particles, *A* and *B*. Figures 21,*C* and 21,*D* show a moving particle which, at event *p*, emits a flash of light, sending light out in all directions.

So far, we have decided that the Aristotelian view is perhaps not a totally appropriate one, for it mixes what we would like to regard as universal structure, intrinsic to space-time itself, with what we would like to regard as structure particular to individual observers. The first step in separating the two types of structure has already been carried out. We saw how to translate between the space-time diagrams as drawn by two Aristo-

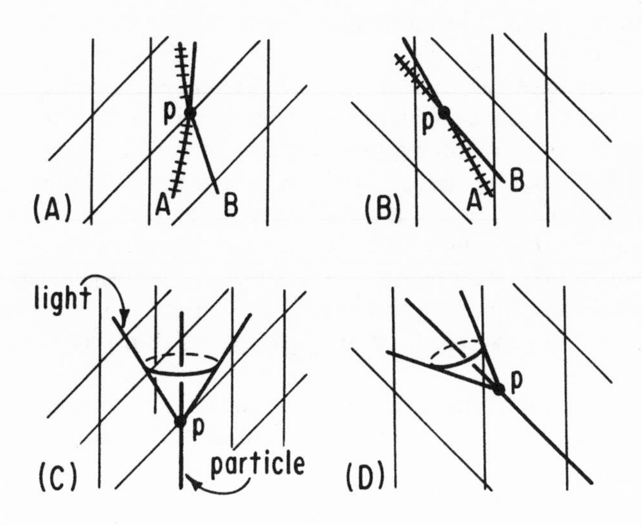

Fig. 21
Two examples of translating space-time diagrams from one Aristotelian setup to another. Figures (*A*) and (*B*) show the collision of two particles; (*C*) and (*D*) the emission of a flash of light in all directions.

telian setups coexisting in the physical world. We now complete this separation process.

Although earlier in this chapter we considered only two Aristotelian setups, we could just as well have considered many. We could have had, for example, seven different Aristotelian groups, each with its badges and watches, all moving—each with constant velocity with respect to the others—through the same region of space-time. Then, for any given physical phenomenon, we could draw seven different space-time pictures. Any two of these pictures would, of course, be related by the operation described earlier in this chapter. In short, one can imagine, within the physical world, an infinite variety of Aristotelian setups, each with its badges and watches. Each describes an event in terms of a position in space and a time of occurrence. Each draws in space-time 3-surfaces to represent "times" and lines to represent "positions in space." Everybody draws the same 3-surfaces. Each group, however, has its own personal set of lines. Each group draws a picture with its own "position lines" as vertical, in which picture the position lines of the other groups are all parallel and straight, but not vertical.

There is really nothing to distinguish any one of these competing Aristotelian groups from any other. Their differing pictures of the world are in some sense all equally valid. This, indeed, was the point of the discussion at the end of chapter 2. Although each may perhaps feel that his own assignments of positions and times are "correct," and that the others are just confused, there seems to be no objective basis for such a judgment.

The Galilean view can be described as a mechanism for ending this bickering between the various Aristotelian groups. By fiat we simply democratize the situation: all the Aristotelian groups are admitted on an equal footing. What structural features result? Each Aristotelian group draws its own time-surfaces—a certain family of three-dimensional surfaces in space-time. But, as we have seen, the various groups all agree on these surfaces. Therefore, no disputes arise here, and this family of surfaces can appear unchanged in the Galilean view. Each

Aristotelian group draws its own position-lines—a certain family of one-dimensional world-lines in space-time. These families, however, are different for the different groups. In the Galilean view, then, all of these various families are admitted simultaneously, with no one family preferred. This, then, is the intrinsic structure of space-time implicit in the Galilean view: There is, in space-time, a certain family of three-dimensional surfaces together with an infinite variety of families of world-lines. More concretely, one can think of space-time according to the Galilean view as a deck of cards (representing the surfaces) pierced with an enormous number of straight wires in every nonhorizontal direction through every point of the deck (fig. 22). An Aristotelian group within space-time is now represented by one particular family of mutually parallel wires (the position-lines of that group). This group would prefer to see the deck so beveled that its lines are vertical; another group, to see it beveled in a different way. The Galilean view simply makes no commitment as to the "state of beveling." The "universal, intrinsic structure of space-time itself," according to this view, is that which refers only to the three-dimensional surfaces and to

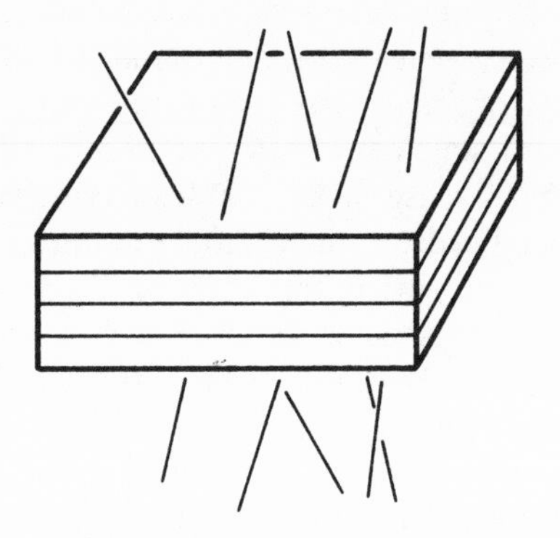

Fig. 22
Space-time, according to the Galilean view. The straight lines represent world-lines of particles, all moving at constant relative velocities. No one family of lines is preferred.

all the lines taken together. The "structure particular to one group" is that which makes reference to one particular, preferred, family of lines.

Two differences between the Aristotelian and Galilean views are immediately apparent. First, whereas the Aristotelian view can more or less be described completely in a single sentence (for example, "An event is naturally characterized by giving its position in space together with the time of its occurrence."), this does not seem to be the case for the Galilean view. Second, whereas a single space-time picture—with its surfaces and lines—seems to adequately and properly represent the Aristotelian view, this also does not seem to be the case for the Galilean view. When we come to actually drawing a picture of space-time, *some* lines must necessarily be vertical at the expense of others: some Aristotelian group will be satisfied, the rest unhappy. This feature, however, is to be regarded as merely an unfortunate difficulty associated with the drawing of pictures, and not as an intrinsic aspect of the Galilean view itself. In looking at a picture representing physical phenomena according to the Galilean view, one is supposed to "ignore verticality."

As we have remarked earlier, the ultimate purpose of these views is in any case the determination of what relationships are available between events—the determination of what "makes sense." We therefore now turn to this question for the Galilean view. The general statement is this: The relationships between events which make sense in the Galilean view will be precisely those which can be determined from the structure on space-time (one family of horizontal 3-planes; an infinite variety of lines) available in this view. In other words, the relationships which make sense in the Galilean view are just those on which all Aristotelian groups will agree. We will give some examples.

"These two events occurred at the same time." This statement makes sense in the Galilean view. In terms of the structure, we interpret "these two events occurred at the same time" to mean geometrically that "these two points of space-time lie on the same horizontal 3-plane." The latter refers to the horizontal 3-planes, that is, to structure available in the Galilean view. In terms of agreement, if one Aristotelian group would claim that "these two events occurred at the same time," then so would all other groups. There is agreement, and so this notion makes sense in the Galilean view.

"The elapsed time between these two events is so many seconds." This makes sense in the Galilean view. It refers to the vertical distance between the corresponding horizontal 3-planes.

"These two events occurred at the same position in space." This does not make sense in the Galilean view. In terms of structure, we do not know which of the many families of lines in the Galilean view are to be used as the "true position-lines." In terms of agreement, one Aristotelian group might answer the question in the affirmative, another in the negative.

"The spatial distance between these two events is so many centimeters." This does not, in general, make sense in the Galilean view. The normal instructions would be to draw one position-line through each of the two points in space-time, and to determine the horizontal displacement between the two lines (fig. 23). The answer will clearly depend, however, on whether the position-lines are those with respect to one or a different family. In terms of agreement, different Aristotelian groups will in general answer this question with different numbers of cen-

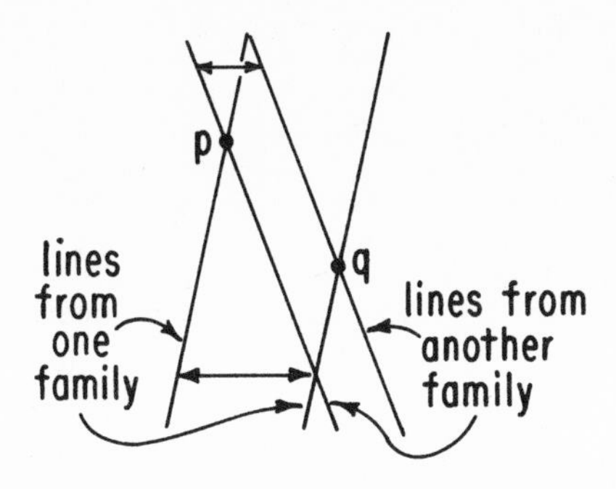

Fig. 23
"Spatial distance between two events" does not make sense in the Galilean view. The value one obtains for the "spatial distance" will in general depend on which Aristotelian group is determining it.

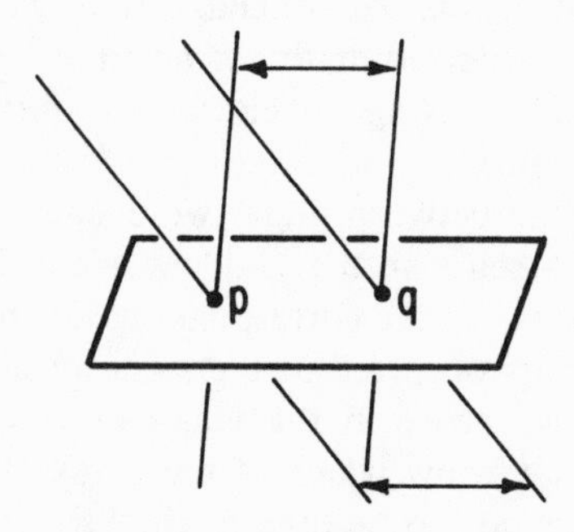

Fig. 24
"Spatial distance" does make sense, even in the Galilean view, for two events which occur at the same time.

timeters (and, indeed, one group would answer "0 centimeters," namely, that group one of whose position lines passes through both points in space-time). There is, in fact, one exceptional situation (fig. 24). Consider two events which occur at the same time (noting that this makes sense in the Galilean view). In this special case, the horizontal displacement between the two lines will be independent of the family from which the lines were chosen (for the horizontal displacement between the lines will always be the horizontal displacement between the two, now "horizontally displaced," points in space-time). Thus, "The spatial distance between these two events, which occurred at the same time, is so many centimeters" does make sense in the Galilean view.

The above summarizes a few of the possible relationships between events which make sense in the Galilean view. But (no matter what one's view is) everything else in the physical world is to be described, ultimately, in terms of events, and in particular statements about other things are to be reduced, ultimately, to statements about relationships between events. Thus, having decided which relationships between events make sense, we can begin to decide which other physical notions in space-time make sense. Some examples follow.

"This particle is moving at constant velocity." This makes sense in the Galilean view, for it becomes, geometrically, "the world-line is straight." To say it another way, the act of beveling the deck to verticalize various position-lines does not affect straightness of a world-line.

"This particle has a speed of 10 cm/sec." This does not make sense. A specific speed refers to elapsed time between two events (which is acceptable), but also to spatial distance (which is not). Indeed, were one Aristotelian group to affirm this statement, another would necessarily say that the particle is at rest.

"This particle is at rest." This does not make sense.

"This particle collided with that one." This makes sense. The statement is that their world-lines have a point in common.

"This rope is straight" makes sense. The corresponding geometrical statement (referring only to Galilean structure) is this: The world-surface of the rope intersects each of the horizontal 3-planes in a straight line segment.

"This particle traveled 10 centimeters" does not make sense.

"Particle *A* is moving faster than *B*" does not make sense. Indeed, one Aristotelian group would see *A* at rest and *B* moving (therefore, faster), while another would see *B* at rest and *A* moving.

"Particle *A* is moving at 10 cm/sec relative to particle *B*." This makes sense. It could be restated thus: On making vertical that family of position-lines which includes the world-line of *B* (so that now, having chosen a particular family, all the Aristotelian notions make sense), the speed of *A* is 10 cm/sec. The point here is that "relative to *B*" allows us to use the world-line of *B* explicitly to single out a particular Aristotelian group. The Galilean view by itself does not rule out such an activity. Rather, it only requires that any singling out that takes place be done only in response to phenomena which are actually available in the statement being analyzed. What is prohibited by the Galilean view is arbitrarily selecting one Aristotelian group for special treatment.

Earlier we gave two examples of laws of physics, and discussed the question of whether they made sense in the Aristotelian view. We will now consider these same two laws from the Galilean view. The Newtonian law of gravitation does make sense in the Galilean view. Indeed, the reference to "one instant of time" poses no problems, for we have, in the Galilean view, our family of horizontal 3-planes to represent the "instants of time." The reference to "distance between (the bodies)" is a bit more subtle. Although it is certainly true that the distance between two events in general does not make sense in the Galilean view, this particular law refers only to the distance between the bodies at that one instant of time. In other words, we only require the notion "distance between two events which occur at the same time," which does make sense in the Galilean view. The law of light refers to the speed of a pulse of light. Since a specific speed for a "particle" does not make sense in the Galilean view, this law also does not make sense.

The Galilean view, although it takes a little getting used to, is really a simple and remarkably natural attitude about how space and time operate. It would be easy to be lulled into the position that it represents the only reasonable attitude one could take.

Finally, we observe that the transition from the Aristotelian to the Galilean view results in fewer things making sense. Although the evidence on this is perhaps a bit scanty, it seems to be the case that physics, at least in its fundamental aspects, always moves in this one direction. For example, relativity theory consists essentially of the realization that certain of the notions in the Galilean view don't make sense either, as we shall see very shortly. In quantum mechanics, to take another example, such notions as "the position of a particle" or "the speed of a particle" do not make sense. It may not be a bad rule of thumb to judge the importance of a new set of ideas in physics by the criterion of how many of the notions and relations that one feels to be necessary one is forced to give up.

4

Difficulties with the Galilean View

The Galilean view seems so simple and so natural that it would be easy to become enamored with it. In fact, this was the state of affairs through a long portion of the history of physics. Eventually, however, there began to be accumulated a number of disquieting observations, observations which apparently could not be properly accounted for within this view. Despite strenuous efforts to save the Galilean view, the pressure from these observations finally became so great that some new view was definitely called for. The result was relativity. It is our intention in this chapter to merely illustrate this struggle by means of two examples.

Let there be constructed two guns, each containing a spring which can be compressed a certain amount such that when released the spring shoots a small steel pellet. The two guns are to be identical, and they are to shoot in the same way every time. Using these guns, we now perform the following experiment. Two people, A and B, each moving at uniform speed carrying a gun, pass each other. At the event of their meeting (p in fig. 25), each person fires his gun in the x-direction, as shown. The resulting space-time diagram, we claim, will be just that of figure 25. In particular, we claim, the world-line of the pellet from A's gun will be different from the world-line of the pellet from B's. The physical reason for this (say, according to A) is "B was moving by me when the pellets were shot, and so the speed of B's pellet will be greater than the speed of my

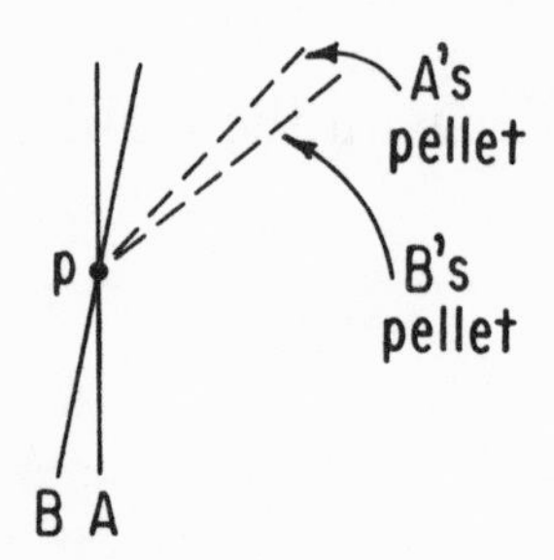

Fig. 25
Space-time diagram for a pellet-shooting experiment. Individuals A and B meet at event p, at which event each shoots a small pellet, using identical spring guns.

pellet, since his pellet acquires an additional speed due to his motion." Now none of this is very shocking. It just says that pellets, once fired, move, not according to the environment in which they find themselves, but rather according to what the shooter was doing, that is, even though the two pellets started at the same event, they describe different world-lines because of the difference between A and B.

We now wish to repeat the experiment above, but with one small change. Instead of guns and pellets, we use flashlights (with directional beams) and pulses of light. Again, as A and B pass, each flashes his light in the x-direction. (Colored filters over the flashlights can be used to keep the two flashes separate.) We ask: Will the space-time diagram in this case look like that of figure 25, or will it look like that of figure 26? Will the two light-beams have the same world-line, or will they be different? Now this is a genuine physical question. It is unlikely to be settled by just thinking about it, but, at least in principle, could presumably be settled by an experiment. I think that all of us would have had the very strong suspicion that the answer would turn out to be that of figure 25.

The experiment is clearly a difficult one to perform in a laboratory, for the light travels so quickly that both beams will reach the far wall of the lab almost immediately, and it will be difficult to determine which reached that wall first. Remarkably enough, it turns out that there is a rather simple experiment to settle this question. There is seen in our universe what are called double-star systems—systems of two stars that orbit about each other. For some such systems, we see the light as emitted from both of the stars, as illustrated in the space-time diagram of figure 27. Consider two beams of emitted light, as shown in the diagram. The emission of light from one star, event *p*, occurs while that star is moving toward us, while the emission from the other star, event *q*, occurs while that star is moving away. In this diagram, the light-beams have been drawn as though figure 25 were the answer to our earlier question. Thus, in this figure the two light-beams reach us ("are seen," event *s*) at the same time. What, then, will be the appearance of the double-star system to us? This is easy to answer. We just draw in the light-beams corresponding to the emission of light at other instants. Then, to decide what we see at any instant (event on our world-line), we find out what light-beams are reaching our world-line at that event. (Note how we deal with questions of this type. We do not suppose that "seeing" is some mysterious mechanism which automatically reaches out and extends our consciousness over space-time. Rather, we draw in

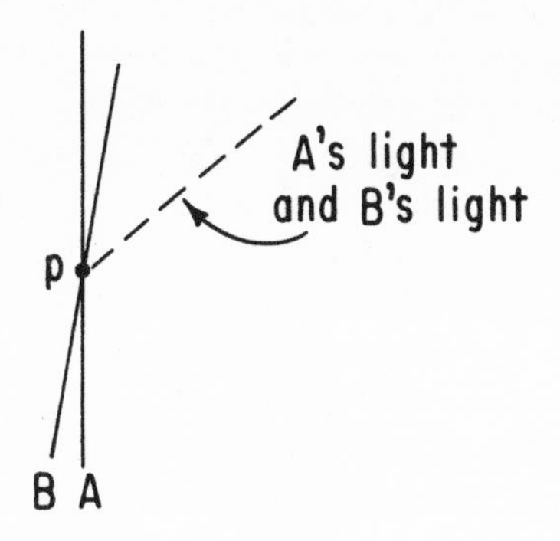

Fig. 26
Figure 25, but now with pulses of light replacing the pellets.

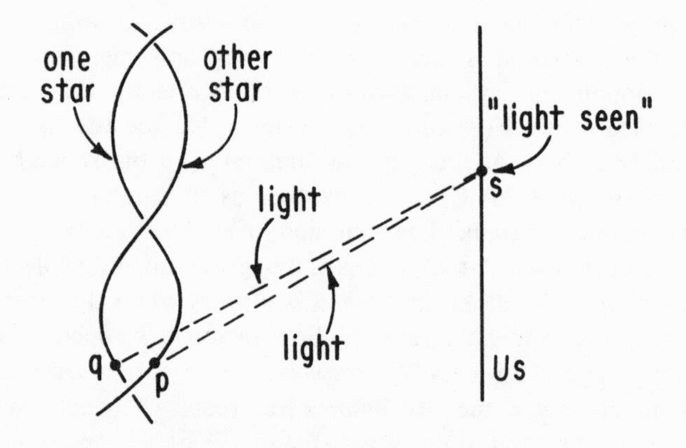

Fig. 27
An experiment to determine whether light behaves as in figure 25 or figure 26. Light from a double-star system is seen by a distant observer (us). This is what the space-time diagram would look like according to the light-propagation of figure 26.

the light-rays, and find the events "receipt of those light-rays on our world-line.") It is also clear what the answer will be. If the situation is as depicted in figure 27, then we shall just see two stars orbiting about each other. Now, however, let us consider the alternative situation, in which the behavior of light is that shown in figure 25 rather than in figure 26. Then figure 27 would be replaced by figure 28. (The physical difference between the two, of course, is that now the two light-beams travel at different speeds, corresponding to the different speeds of the emitting stars at the events of emission.) Now, although the two light-beams were emitted simultaneously (events p and q, which lie on a horizontal 3-plane), they are received at different times (events r and s). Thus, what we will see at any one instant (event, on our world-line) will be light emitted from one star at one time and from the other star at some quite different time. Were this the situation, then, the stars would appear to execute very complicated motions, since we are really looking at different stars at different times.

Thus, the experiment is this. One looks at a double-star system, and asks "Do the stars appear to be just orbiting around each other, or do they appear to be executing some complicated motion?" The result of observations is that it is the former. We conclude, therefore, that figure 27 correctly represents the stars, light, and us, and therefore that figure 26 represents the probable result of the experiment performed with flashlights.

This, it seems to me, is at least a mildly surprising conclusion. There was no particular reason to suspect that light would behave so differently from steel pellets. (Well, maybe there was a little hint. One often sees slow-moving steel pellets, but never slow light.) In any case, this seems to be the way light is, and we shall have to live with it. It is as though light, once released from its flashlight or star, decides what its world-line in space-time will be by reference to space-time itself and that alone—without any reference to the behavior of the emitter. (Light, apparently, is like a fish swimming in the ocean, which moves relative to just the ocean, no matter how the person who released the fish was swimming.)

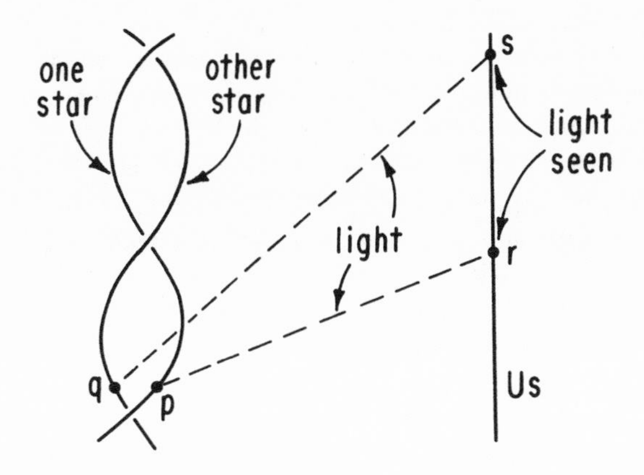

Fig. 28
An experiment to determine whether light behaves as in figure 25 or figure 26. This is what the space-time diagram would look like according to the light-propagation of figure 25.

So far, we have only discovered that light has a certain intimacy with space-time; there is nothing immediately damaging to the Galilean view in this. One thing, however (we still believe the Galilean view), is certain. The following statement could hardly be true: Light-pulses travel at 3×10^{10} cm/sec. This statement, in fact, makes no sense in the Galilean view. Furthermore, we can now even be certain that the following statement cannot be true: Light-pulses travel at 3×10^{10} cm/sec relative to the emitter of the light. In the case of the second, the reason is the following. We have just seen that light, once released, moves along its own world-line, no matter who emitted it. Hence, if light that I emit always moves at 3×10^{10} cm/sec relative to me, then light that anyone emits must move at 3×10^{10} cm/sec relative to me (for light doesn't care who emitted it). Therefore, light which anyone emits must move at 3×10^{10} cm/sec relative to anyone. But this is the content of the first statement, which we decided made no sense. Thus, the second statement (which, together with our observation above about light, implies the first) can make no sense. Our conclusion: It is false that light always moves at 3×10^{10} cm/sec relative to the emitter of the light.

We, of course, are perfectly free to make clever arguments and conclude anything we want. Nature doesn't care what we conclude. Our "conclusion" above is in any case a genuine statement about the physical world. It can be tested experimentally. One in essence asks a number of observers to measure the speed of light, and finds out what answers they obtain. (The actual experiment is technically rather more complicated than this, but its details don't matter.) The result is that every single one of our observers reports back "I found the speed of light to be 3×10^{10} cm/sec."

The discussion above is a good example of the kind of activity that goes on in physics. One more or less convinces oneself of a view of the structure of space-time (Galilean), and makes a few observations about how light behaves (the double-star experiment). From this, one draws conclusions about how light should behave in other circumstances (speed not always $3 \times$

10^{10} cm/sec). Light refuses to behave as we have concluded it should. We face a problem, for we really don't understand what is going on. (Of course, at the time nobody would have suspected that the Galilean view was the weak link, for it always sat inconspicuously in the background, dominating one's thinking while not seeming worth examination.) We will postpone for a moment our discussion of possible reactions to this problem.

Our second example of a possible difficulty with the Galilean view involves certain elementary particles called mu-mesons. These particles can be produced in the laboratory and, when produced, are found to decay (in more detail, to explode into one electron and two neutrinos) about 10^{-6} sec (1 microsecond) after they are produced. (It turns out, incidentally, that the mechanism for this decay, by what is called the weak interaction, is rather well understood.) These mu-mesons are also produced naturally outside of the laboratory, in the following way. Cosmic rays (fast-moving nuclei from various atoms) constantly impinge on the earth from outer space. (Exactly how these cosmic rays got there, incidentally, is perhaps not so well understood.) In any case, when a cosmic ray comes toward the earth, it normally strikes an atom of air in the upper atmosphere. The result of this collision is that numerous elementary particles of all kinds are produced. Because of the high speed of the incoming cosmic ray, these particles "shower" down on the earth. Among all these elementary particles are mu-mesons. Using particle detectors on the surface of the earth, one can detect the presence of these particles, and in particular of the mu-mesons. Now, one can estimate how high in the atmosphere the collision took place (event p in fig. 29), and one can determine approximately how fast the mu-meson is traveling when it reaches the earth. From this data, one can compute how long it took between the production of the mu-meson and its detection at the surface of the earth. The result of this calculation? About 10×10^{-6} sec (10 microseconds). Question: Since mu-mesons apparently can only survive 1 microsecond before they

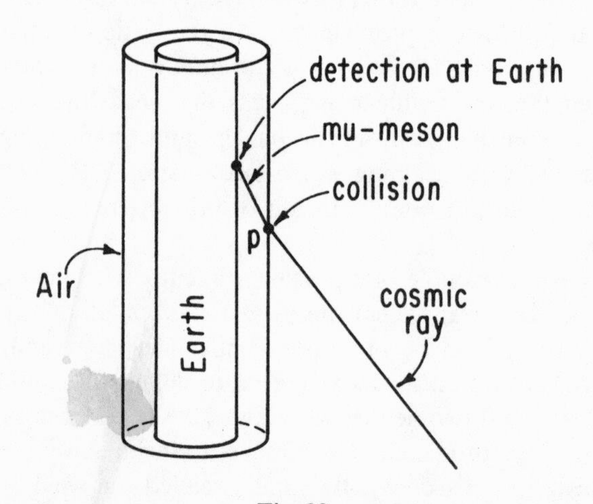

Fig. 29
Space-time diagram for a mu-meson experiment. A cosmic ray collides with the atmosphere at event p, releasing a mu-meson, which is finally detected at the surface of the earth.

decay, how do these mu-mesons manage to make it to the earth without decaying, since this journey takes 10 microseconds?

What this question is really asking is what is it about the mu-mesons from cosmic rays that causes them to live longer (or, at least, apparently live longer) than the mu-mesons produced in the laboratory? Could it be that there are actually two types of mu-mesons, longer-lived ones from cosmic rays and shorter-lived ones from the laboratory? As far as one can tell, they are identical in every other respect. Could it be that mu-mesons just live longer in air than they do in a vacuum? Laboratory mu-mesons, placed in the air, retain their short lives. Could the cosmic-ray collision actually take place lower in the atmosphere than we think? One can send balloons with particle detectors successively higher into the atmosphere to determine where the collisions take place. Could it be that the cosmic-ray mu-mesons live longer because they are moving much faster than the mu-mesons produced in the laboratory (which they are)? One can test this as follows. One produces in the laboratory mu-mesons moving at various speeds, and determines whether the lifetime of the mu-meson depends on the speed at which it is moving. The result of such an experiment is shown in figure 30. The vertical axis is the "lifetime of the mu-meson," and the horizontal axis is the "speed of the mu-meson." Each mu-meson produced has a measurable speed and a measurable lifetime, and so results in a single point on the graph. Producing many mu-mesons, we obtain many points, which, as it turns out, all lie on the curve shown. If all mu-mesons had the same life, the curve would instead have been the dashed horizontal line. This experiment, then, apparently answers our question quite satisfactorily: Slow-moving mu-mesons have a short lifetime (10^{-6} sec), faster-moving mu-mesons a longer lifetime, and so on. (Note, from the graph, that no mu-mesons move faster than 3×10^{10} cm/sec, a point to which we shall return later.) The mu-mesons from cosmic rays have a long lifetime because they are moving quickly.

We seem, then, to be pushed toward the following as a law of physics: The faster a mu-meson is moving, the longer its lifetime. This, however, is crazy, for this supposed "law of physics"

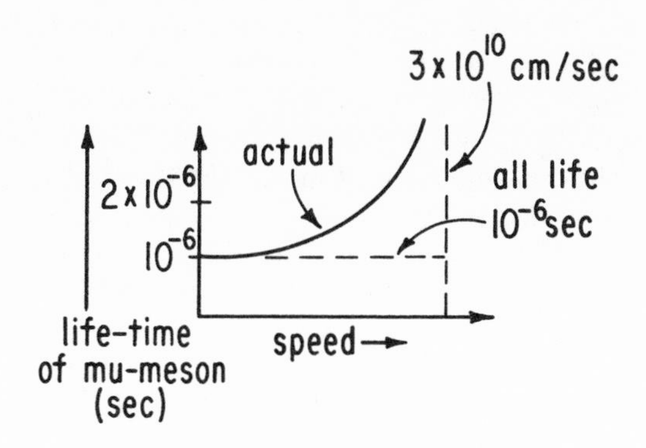

Fig. 30
Graph of the lifetime of a mu-meson as a function of its speed.

makes no sense in the Galilean view (for, although "lifetime" makes sense there, "faster-moving" does not). Here, then, is our second problem: Nature seems to obey a law which we deem not sensible.

Confronted by these two problems (and, historically, by others), how should one react? One possibility would be to essentially try to ignore them by some stance along the following lines: "One generally accepts the Galilean view of space-time structure, but one must remember to be very careful about light-pulses and mu-mesons, for they have special properties which have to be taken into account." This, from the point of view of physics, would be a terrible resolution. Physics is not concerned solely with observing nature and cataloging these observations in large volumes. In addition, one would like to have some feeling of understanding nature, in, among others, the following sense. One would like to feel that he can, with some confidence, make statements which apply, not just to those special situations in which experiments have been performed, but also to similar situations in which the experiments have not yet been performed. That is, one wishes to formulate statements about nature which have some generality and some predictive power. The statement at the beginning of this paragraph fails completely on these counts. It does not suggest, for example, whether pi-mesons will behave as mu-mesons, or whether they will behave "normally." It does not suggest how, in the case of light, two Aristotelian groups observing the same pulse of light can both claim it moving at 3×10^{10} cm/sec.

So, we have decided that these two problems cannot be ignored (a decision which would come easily and naturally to physicists, in whom the attitude of the previous paragraph is quite ingrained). The next tack one might take would be to look for the simplest, the most mundane, the most conservative explanation for the observations one could find. In physics, as a general rule, one does not jump to more complicated explanations unless absolutely forced to do so. (What do we mean here by an "explanation"? I find this question difficult to answer, but, if pushed, might say "a list of statements about the way nature behaves which is reasonably coherent and which gives one a feeling that he has a comprehensive picture." A more honest, but less helpful, answer might be "something is an explanation if physicists generally agree that it is." This whole issue, which perhaps lies between nature and sociology, seems to be a bit vague. Quite possibly, an attempt to make the word *explanation* more precise may do more harm to the field than good.) In any case, we look for mundane explanations. We have already discussed, preceding figure 30, some examples in the case of mu-mesons. As we discussed there, one tries, for each explanation, to subject it to some sort of observational test to see if it works. As we remarked there, none of these proposed explanations did seem to work. Similar remarks would apply to the problem involving light. One might, for example, try to invent an explanation based on the fact that one of the two experiments (the one with the double-star system) involves light traveling through empty space, while the other (different observers measuring the speed of light) involves light in air. Such explanations, again, are worked around to the point at which they can be tested in some way. Again, they did not seem to work.

The process described in the previous paragraph went on for many years—first the most mundane explanations, then the less mundane, and so on. It was eventually realized that the problem might lie with the Galilean view itself, that some new view of space-time might provide an explanation. This "new view" is, of course, relativity.

B

General Relativity

5

The Interval: The Fundamental Geometrical Object

We decided (or at least asserted) in chapter 4 that the Galilean view needs modification. We will now set down the basic ground rules.

We have already gotten into difficulties once with the Aristotelian view. We rushed in too quickly, allowing ourselves to use any sort of instruments or ideas with abandon; we imposed structure right and left, finding out only later (with the Galilean view) that some of it wasn't "intrinsic to space-time." It is tempting to try to avoid making the same mistakes again by adopting an ultracautious attitude. We shall be highly skeptical, not accepting anything until it is soundly grounded in many observations; we shall proceed very slowly and very carefully in very small steps. Such an attitude (unfortunately, if you wish) seems never to get one anywhere. Had it been adopted at the beginning of this century, we would probably not have relativity today. What is needed, experience has shown, is the right mix of caution and daring. One must at times be rash, accepting (perhaps temporarily) ideas with very little observational basis; one must at other times be ultracautious, examining "obvious" notions with care. The art (and it is an art) consists of making judicious choices of what is to be in the first category and what in the second. It is these "judicious choices," then, that we must now make.

I cannot hope to make what ultimately turned out to be the "right" choices seem natural and reasonable. There is, to put it dramatically, an imaginative leap grounded in judgment and experience and little else. The choice, however, is this: We shall retain the broad, qualitative features of space-time. That is to say, we shall retain the notion of an event, and the idea of assembling those events into space-time. Points still represent events, world-lines still represent particles, two-dimensional surfaces still represent ropes, and so on. Intersections of world-lines still represent collisions of the particles, a world-line meeting the world-region of the earth still represents the meeting of the particle and the earth, a world-line on a two-dimensional surface still represents a bead sliding on a wire. All else, for the moment at least, we discard. We don't have clocks, we don't have "spatial distance between two events," we don't have "events occurred at the same time," we don't have "speed of the particle," and we certainly don't have any Aristotelian setups. In short, we retain the general idea of space-time as the way to represent the world, and discard all the detailed, geometrical, numerical information. In the transition from the Aristotelian to the Galilean view, we switched from a rigid picture of space-time to a "sliding-pack-of-cards" picture. We are now going to the extreme in this direction. It is as though space-time were drawn on a rubber sheet, which can be stretched, pulled, and bent. All that we retain, all that we care about for the moment, are the broad, qualitative features of the various points, lines, surfaces, and so on drawn on the sheet. (True, any actual drawing of space-time will involve geometrical relationships that we do not wish to regard as meaningful at this point. It's the same problem as that in the passage from Aristotelian to Galilean—and the solution is the same. One must just get used to viewing space-time diagrams as though they were printed on a rubber sheet, subject to being stretched and pulled without changing anything.)

The choices of the previous paragraph do not sound as revolutionary as they were because we have prejudiced the situation by our treatment, in the first four chapters, of prerelativistic space-time. Even so, the choices, I would like to claim, are a bit shocking. Originally, recall, events, space-time, and so on were introduced merely as an aid in picturing what we thought of as the "real geometrical structure" of space and time. These notions were just a convention we all agreed to adopt to communicate with each other. Now, however, these notions are brought to the fore, with the (previously) "more natural" ideas, such as spatial distances and elapsed times, suppressed.

Shall we just stop here and call it a "view"? It would certainly be a very simple view—one almost immune from any conflicts with observations. (I would be hard-pressed even to imagine any observation which wouldn't fit into this broad framework.) Alas, we cannot. The problem is precisely that what we have so far is so immune from conflicts with observations. Indeed, it doesn't assert much of anything about the world at all. We all have at least some intuitive notions about spatial distances, elapsed times, and so on, and these must, in some way or other, be brought into our view if it is to be other than empty. A theory of physics in general, and a world-view in particular, must make at least some commitment about the physical world if it is to be worth anything at all.

The conclusion, then, is that we must somehow work into this broad framework some hard, numerical, geometrical information about space-time. We need something to replace the old spatial distances, elapsed times, and so forth. Here is where we wish to be ultracautious. The problem (it is now known) is that we were entirely too cavalier in describing, for example, the original Aristotelian setup. We allowed ourselves "watches" without further discussion, and we implicitly assumed a number of properties of these "watches" (some of which, as we shall see shortly, are just not true in our world). We allowed ourselves "meter sticks" (used to measure the distances recorded on the badges) with implicitly assumed properties. We in effect allowed ourselves to "just know" about space-time around us without a careful physical prescription for what is to be done.

This time, we wish to be more careful, in at least the following senses. First, we shall not allow ourselves to use any old measuring instrument we happen to stumble across; rather, we shall restrict consideration to some minimum number. Second, we shall require that these instruments—whichever ones we select—shall first be properly represented within our space-time framework. We do not accept instruments that we don't even understand well enough to draw in space-time. Third, we require a clear, explicit statement of those properties that our instruments (now represented in space-time) are to be assumed to have. We may, of course, suppose what properties we want and need, but we do require that such suppositions be explicit.

The paragraph above is of little help in deciding what instruments we should allow ourselves. For this, we must again try to make a careful judgment (translated, lucky guess). After mulling the matter over a bit, one might hit on the following idea. What sorts of instruments are actually needed to obtain geometrical information about space-time? One needs essentially two different types of instruments. One of these instruments must go out into space-time (that is, away from our world-line), collect or respond to what space-time is like, and carry that information back to us. Without such an instrument, we should forever be confined, in terms of our knowledge, to our own world-line. A second instrument would then also be needed to "record" or "make numerical" the information brought back. Without this second instrument we would only be able to sense space-time away from our world-line, not say anything concrete about it. This idea vastly simplifies our search for instruments, for we know not only the number (two), but also, in general terms, what the two instruments are to be doing for us. We now turn to the specific choice of instruments.

What shall we use for the instrument which reacts to space-time away from our world-line and brings information back? Let's try a few possibilities. Consider first a meter stick. It's a whole meter long, and so it certainly "sticks out into space-time away from our world-line." There are, however, two unpleasant features about meter sticks. The first is that, in terms of space-time, a meter stick is a rather complicated object: It's represented by a two-dimensional surface. (Of course, the markings on the meter stick would be lines ruled on this surface.) The second feature is perhaps even worse. How do we go about

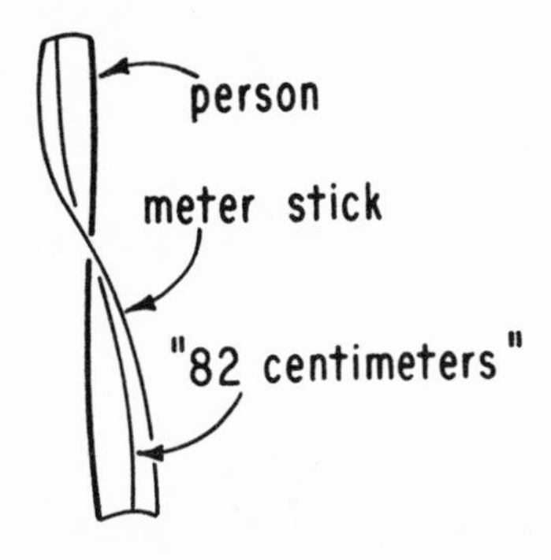

Fig. 31
Space-time diagram of a meter stick.

reading the marks on the meter stick? The world-line, say, of the mark "82 centimeters" on the meter stick will be out in space-time away from our own world-line (fig. 31). If we are going to invent some mechanism to tell us where that mark is, then we might as well use that mechanism to tell us about space-time. (It will not do, of course, to go over to the "82-centimeter" mark to see where it is, because if our world-line gets over there then we are there, and so we are directly reading things on our own world-line, without the need of any meter stick.) Meter sticks, in short, are not an entirely satisfactory way to determine what space-time is like. (Recall, now, their extensive use for the Aristotelian setup.) Consider, then, the possibility of using particles (say, thrown by the observer) to react to space-time. One could arrange for the particles to be bounced back to the observer to tell him what space-time is like away from his world-line (fig. 32). Particles, indeed, are much better than meter sticks for our purposes (and, in fact, they could perfectly well be used, albeit at the expense of somewhat more complicated constructions). The problem with particles is that they can have all sorts of different speeds when thrown out, which is just another complication. The solution, perhaps by now obvious, is to use light-pulses. Light will certainly go out

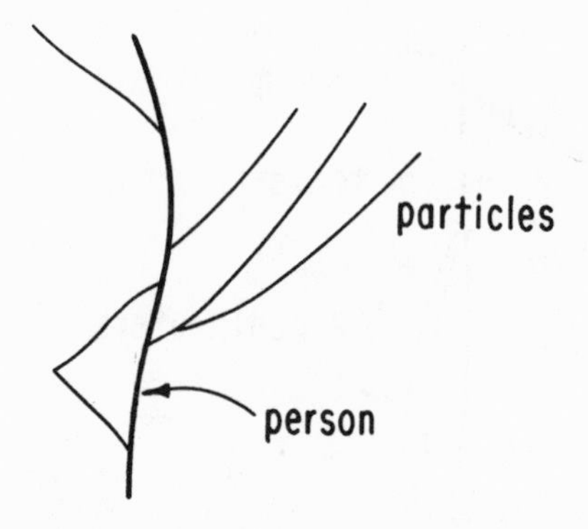

Fig. 32
Space-time diagram of individual using particles to detect the structure of space-time away from his world-line.

from our world-line, and it will certainly react in some way to what is happening in space-time. In fact, as we saw earlier ("light, once emitted, moves within the environment of space-time, independently of what the emitter was doing"), light reacts to space-time rather well. By mirrors, for example, one can arrange for the light to get back to us to tell us what space-time is like. The advantage of light over particles is that you have to decide how fast to throw a particle (that is, one aspect of the instrument projecting the particle will be "special to the observer" rather than "intrinsic to space-time"), whereas no such decision is required for light.

We have now settled on our instrument which "probes space-time." What about the one which "records"? Again, a meter stick (with all those marks on it) comes to mind, but our previous objections remain in force. After contemplating various other possibilities, one might eventually decide on clocks. They have an obvious advantage over meter sticks, in that they stay on one's world-line (assuming that the clock is held, and not thrown off), and thus can be read locally without additional constructs (that is, we avoid the earlier problem of the "82-centimeter" mark). Further, a clock doesn't have any extraneous features; it doesn't do anything except produce numbers, the "times." A clock is a perfect "pure recorder."

To summarize, we have so far selected our general framework (events, space-time, and so on), and we have selected our two instruments (light, to "reach out," and clocks, to "record"). The next step in our program is to represent these instruments within space-time.

Light is easy. A pulse of light is described by its world-line. The representation of a clock is somewhat more complicated. First of all, when we say "clock" we are thinking of its being idealized to be very small. Second, all our clocks are supposed to be the same (say, all made on a certain assembly line in Hoboken). Third, when we say "clock," we really mean "clock plus calendar," we wish to ignore the fact that a clock face looks the same every twelve hours. These conventions having been made, how shall we represent clocks? First of all, a clock, being very small ("particle-like") may be assigned a world-line. This single line, however, does not exhaust what a clock does, for it carries no information about the "times" that a clock reads. How shall we represent these "times"? The point is that "the time reading of a clock" is a function which assigns, to each point on the clock's world-line, a number (the "time"). In physical terms, a point on the clock's world-line represents an event occurring in the immediate presence of the clock, and the number assigned to this point represents the reading on the clock's face at the occurrence of this event (fig. 33). This, then, is how we shall represent our idealized clocks: each consists of a world-line to each point of which there is assigned a number. (We shall, needless to say, refer to the number assigned to point p as "the time, according to this clock, at event p.") Note that clocks do not assign times to events not on their world-lines. If we wish to discover "the time" at some other event, we shall either have to run a clock through that event, or else we shall have to somehow carry information from some other clock over to the event in question.

The paragraph above completes the second step in our program. We now turn to the third: explicit statement of the prop-

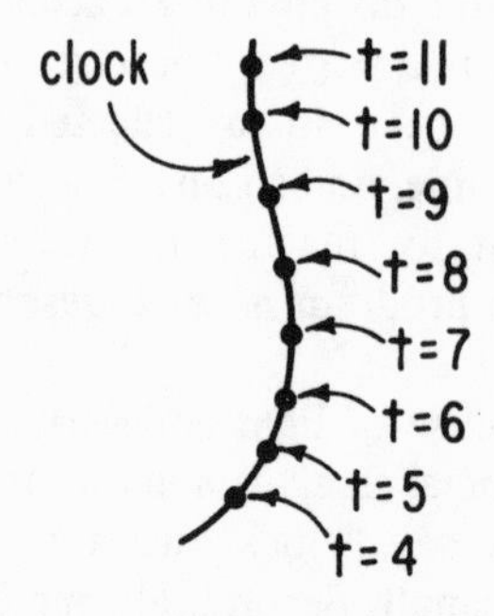

Fig. 33
Representation of a clock in space-time. The clock possesses a world-line, and furthermore assigns a number t, the time-reading of the clock, to each event on that world-line.

erties assumed of our instruments. For light, once again, this is easy. Essentially the only property of light that we shall assume is that whose discussion follows figure 26. We shall suppose that two pulses of light, emitted in the same direction from the same event, move together, no matter what the emitters are doing. This property, after all, is one of the things that makes light attractive for our use. For clocks, once again, the situation is somewhat more complicated. Let us consider first the following property. Let two clocks have coincident world-lines until event p, at which event they are separated (fig. 34). Let the clocks be brought together, to remain together thereafter, at event q. Since p is a point of the world-line of clock A, this clock assigns to p a number, the time according to clock A of p, which we may write as $T_A(p)$; since q is a point of the world-line of clock A, this clock assigns to q a number, $T_A(q)$. The difference between these times, $T_A(q) - T_A(p)$, represents physically the elapsed time, according to clock A, between event p and event q. Similarly, using clock B, we obtain $T_B(q) - T_B(p)$. Shall we now assume of clocks the following property: Under this arrangement, $T_A(q) - T_A(p) = T_B(q) - T_B(p)$? That is to say, shall we assume that all clocks measure precisely the same elapsed time between two fixed events, no matter what world-line the clock traverses between the events? We claim that this would not be a good supposition, for the two reasons given below.

Let us, for a moment, make the supposition above, and see what its consequences would be. Fix, in space-time, some reference event p (say, the signing of the Declaration of Independence), and let us arbitrarily assign to this particular event time $t = 0$ (fig. 35). Consider now any other event q in space-time. We may assign to this event q a "time" as follows. Find a clock, A, which passes through both events, p and q, and thereby determine the numbers $T_A(q)$ and $T_A(p)$. Using this clock A, we then assign to event q "time" $t = T_A(q) - T_A(p)$. (Note that these instructions assign to the reference event p itself time $t = 0$, as we would wish.) This assignment of a time to event q will in fact be the same no matter what clock we have sent from p to q (that is, the same for clock A as for clock B), by our present supposition. Thus, event q is assigned a unique time. But all this is true for any choice of "event q." Thus, to each point in space-time there is assigned a unique time. Having made such assignments, we may now, if we wish, draw in space-time the surfaces corresponding to the various times, that is, we may draw the surface "$t = 3$" (which passes through precisely

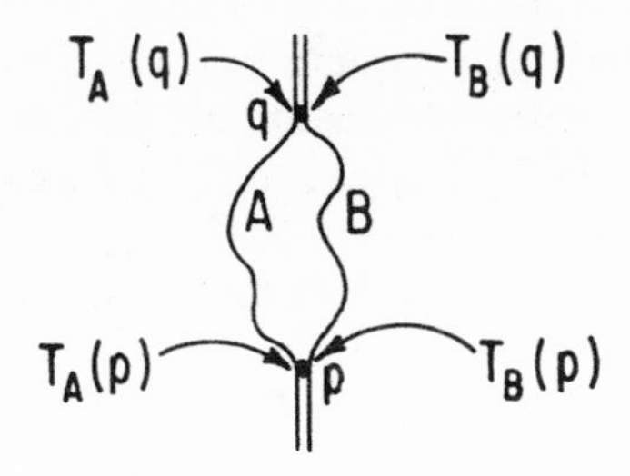

Fig. 34
Space-time diagram of an experiment involving two clocks, A and B. The clocks separate at event p, and return together at event q. Is the elapsed time according to A between p and q the same as that according to B?

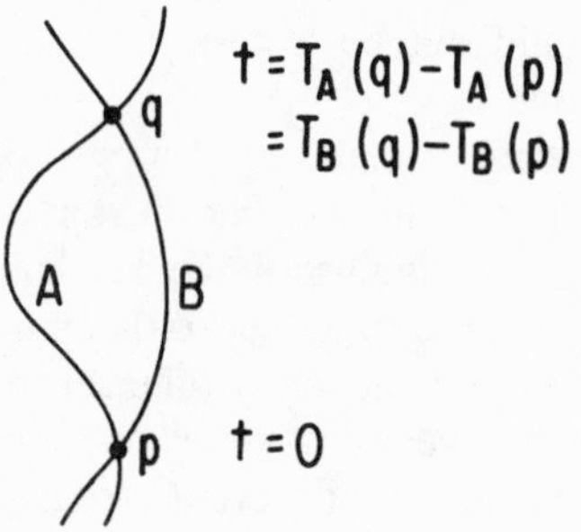

Fig. 35
Under the supposition that the answer to the question of figure 34 is yes, we obtain a universal time in space-time. The measured elapsed time from p to q must be the same no matter what clock measures this time, and so, assigning to p time $t = 0$, a unique time is assigned event q.

those events assigned, by our construction, time $t = 3$), and so on. The result is a collection of time-surfaces in space-time. The point, then, is that our supposition leads immediately to a slicing of space-time by time-surfaces. But we have now, essentially, arrived back at the Galilean view. (Indeed, looking back at our original construction for this view, we see that there we implicitly made the assumption that "clocks, once synchronized, remain synchronized," an assumption which is essentially the supposition now under consideration.) But we do not wish to arrive back at the Galilean view, for, as we have seen, this view leads to a number of problems with observations. The only way, apparently, to avoid the Galilean view is not to make the present supposition.

A second argument against this supposition is observational. In a certain sense, the experiment illustrated in figure 34 has been performed, with negative results. We first note that we may regard a mu-meson as a sort of simplified "clock," in the following sense. A mu-meson apparently "keeps track of the time since it was created," so that, when this time reaches 10^{-6} sec, the mu-meson decays. (Of course, the information given by a mu-meson clock is much less detailed than the information given by one of our clocks, but nonetheless a mu-meson is rather "clocklike.") Now consider again the experiment illustrated by figure 29. We insert, in this experiment, a couple of clocks, one of which remains on the surface of the earth (clock *A*), and another of which goes up into the atmosphere, experiencing the event "collision of the cosmic ray with air molecule," and returning to the earth with the mu-meson (clock *B*) (fig. 36). (This, of course, is a thought-experiment: It would be difficult in practice to cause clock *B* to act as we have suggested.) Now, clocks *A* and *B* are together at event *p*, and also at event *q*. One might expect that the elapsed time according to clock *A* between events *p* and *r* (where *r* is that event on the surface of the earth which Aristotelians would say occurred at the same time as the event of collision, *s*) would be the same as the elapsed time according to clock *B* between *p* and *s*. On the other hand, the elapsed time according to clock *A* between *r* and *q* will be about 10×10^{-6} sec (for that is the content of

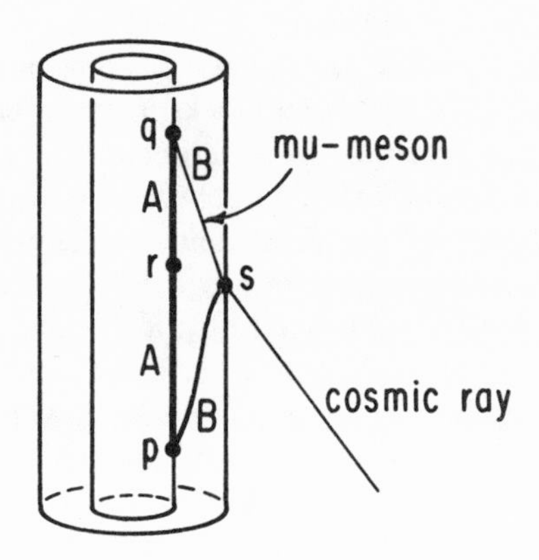

Fig. 36
The mu-meson experiment revisited. Now, however, we imagine introducing two clocks, *A* and *B*. The experimental result with mu-mesons suggests that the answer to the question of figure 34 is no.

the original discussion of the experiment), while the elapsed time according to clock B between s and q will be less than 10^{-6} sec (for the mu-meson does not decay, whence, regarded as a clock, it measures elapsed time less than 10^{-6} sec). Thus, what seems to be the case is that clock B will measure a somewhat shorter elapsed time between events p and q than will clock A (more precisely, at least 9×10^{-6} sec less). Clearly, this experiment is not "hard evidence" on the question of equality of elapsed times, because, for example, it is only in part an actual experiment, the rest thought-experiments and "reasonable assumptions" about the world. Nonetheless, the argument, taken as a whole, does rather suggest that we should not make the above-mentioned supposition.

For these two reasons, then, we decide not to suppose that elapsed times between two fixed events, as measured by clocks with different world-lines between those events, will always agree.

If we cannot have that property, what about the following one? Let the two clocks, A and B, describe the same world-line between events p and q (fig. 37). Again, we can determine the T_As and the T_Bs, and we can ask whether or not $T_A(q) - T_A(p) = T_B(q) - T_B(p)$. Our clocks, recall, were constructed to be identical. In the present experiment, the two clocks are following an identical world-line between the events. Does not equality therefore hold by "identicality"? The answer, it seems to me, is "not necessarily." Although it is certainly true that clocks A and B were constructed identically, and although it is also true that the two clocks follow the same world-line between p and q (in the present issue), the clocks have been permitted to have quite different histories before they even reached event p. What we are here asking, in essence, is whether or not the past history of a clock (before p) affects its "ticking rate." One could certainly imagine that the answer could go either way. Since we really should make some commitment on this issue, and since there are no compelling reasons to suppose otherwise, let us suppose that at least this last property of clocks will hold. (There is, additionally, some indirect experimental evidence suggesting this supposition.) This, then, is the property we are

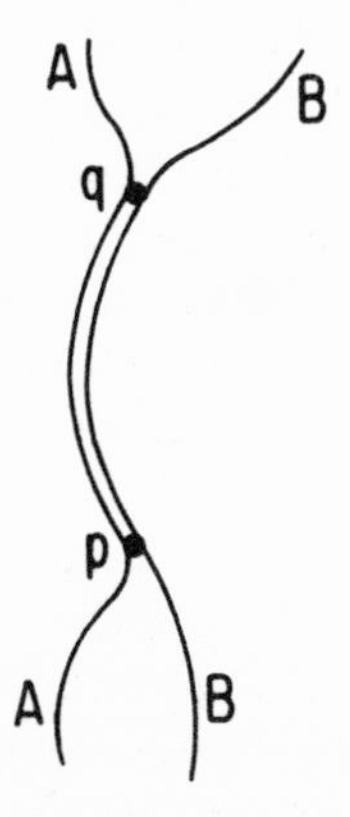

Fig. 37
A second experiment involving two clocks. Now, the clocks are together between events p and q. Will they measure the same elapsed time between event p and event q?

supposing clocks to have: Two clocks having identical world-lines between two fixed events measure the same elapsed times between those events.

We emphasize that either of the two properties of clocks discussed above could, in principle, have gone either way. These are genuine physical questions about the way nature behaves. As far as I can see, there are no compelling, a priori reasons why the answers in either case should be one way or the other. The experimental evidence is perhaps a bit thin. What we must do is take what evidence we can find, together with our prejudices and guesses, and make a kind of working assumption to allow us to get on with it. Such "working assumptions" can (and often do) turn out to be simply not realized in our world. Were this to happen in the present case, we would of course have to come back to this point and try to discover some new collection of working assumptions and some new view of space-time.

The upshot of our property of clocks is this. Given any world-line of a particle, one acquires an assignment of times to points of that world-line. (Physically, the assignment is obtained by carrying a clock alongside the particle, and using the readings of the clock to obtain the "times.") This assignment of "times" to the points of the world-line of any particle is unique as far as time differences are concerned. (Here is where we are using our property of clocks. When we say "carry a clock alongside the particle" we do not specify the past history of the clock. We are thus assuming implicitly that ticking rates are independent of past history.) To summarize, world-lines of particles now acquire time-functions.

So far, we have decided on the general framework (events, space-time, world-lines, and so on), on the instruments to be used in determining geometrical information about space-time (light-pulses, clocks), on the description of these instruments in space-time (light-pulses are world-lines, clocks are world-lines to each point of which there is assigned a time), and on the properties to be assumed for these instruments (that the world-line of emitted light is independent of the emitter, that two clocks having the same world-line between two fixed events measure the same elapsed times). However, what we are ultimately after is none of these things, but rather relationships between events. The whole idea, remember, is that, once everything in the world has been described in terms of events, relationships between events yield relationships between things in the world, that is, physical statements about how the world operates. We turn, now, to the use of our instruments and framework in the determination of relationships between events. We emphasize again that we are permitted, in this determination, to use only our two instruments, their description and properties.

Fix two events, p and q, in space-time. We wish to construct, using our instruments, some relationship between them. How shall we proceed? One might, as a first step, think of sending a light-pulse between the two events. This, however, will not work, for in general, given events p and q, there will be no light-pulse between them. (Indeed, as we saw earlier, the set of events reached by a light-pulse from p, the light-cone of p, is only a certain collection of events in space-time, and may not include the event q.) One might think of sending a clock between the two events. However, the same problem arises. In general, there may be no clock whose world-line meets both events. What shall we do? Recall the original purpose for the two instruments. Clocks were supposed to record information and light-pulses to carry information around. This suggests that we should use both a clock and light-pulses, and that we should keep the clock around one of the events (say, p), using light-pulses to respond to event q and bring to us (traveling with the clock) information about it.

Let us consider, then, the following arrangement (fig. 38). We send one of our standard clocks through event p. We

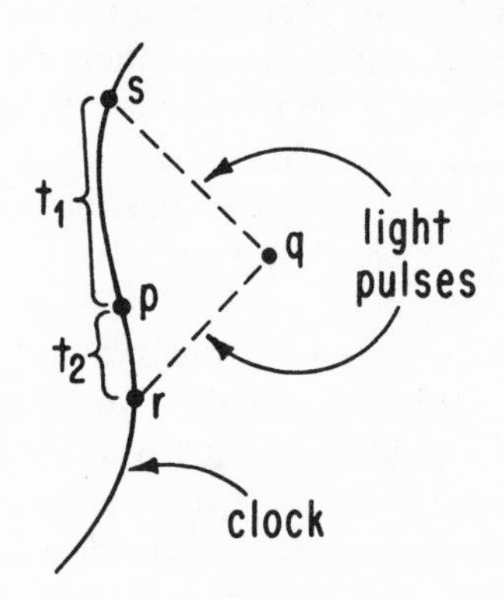

Fig. 38
The construction which expresses numerically the relationship between two events, p and q. A clock is allowed to experience event p, while a light-pulse, leaving the clock at r, reaches q, and another pulse, leaving q, returns to the clock at event s. The clock then measures two elapsed times, t_1 and t_2.

send a light-pulse from the world-line of that clock (say, emitted at event r) to event q. A second light-pulse is to be emitted from event q so as to again reach the world-line of the clock (say, arriving at event s). The point is that we (who travel with the clock, and who experience event p) cannot thereby experience event q. We therefore allow the light-pulses to experience q, and we record the events of emission and reception of the light. The two events r and s (on our world-line) represent "information about q as carried back to us by light." Now the purpose of the clock was to make information numerical. What numbers are available to us from this arrangement? Of course, all a clock will be willing to do for us is assign times to points on its world-line. In this case, we have three such points, r, p, and s. Thus, using the clock we can measure two numbers: the elapsed time between r and p according to the clock, $t_2 = T_A(p) - T_A(r)$, and the elapsed time between p and s according to the clock, $t_1 = T_A(s) - T_A(p)$. In this way we obtain, from the two events p and q, two numbers, t_1 and t_2. These numbers in some sense or other describe a relationship between the two events. Note that we have, above, used the instruments for the purposes for which they were intended. I would also like to claim that the construction above is perhaps the simplest which yields numbers from two events according to our general program.

It is useful, before we begin worrying about what we shall do with t_1 and t_2, to get a feeling for what they mean physically. Let us therefore imagine someone who travels along with the clock, and who has only an everyday notion of distances, speeds, and times. He knows nothing of space-time, various views, relativity, and so on. Let us determine how he would describe this experiment. He would, of course, say, "At r I emitted light toward q. Then p occurred. Still later, at s, I received the return light signal from q." Let us now imagine asking this person the following question: "What was the spatial distance between p and q?" He might answer as follows. "Well, I sent out the light at r, and then received it back at s. Since the elapsed time between r and p is t_2 sec and that between p and s is t_1 sec, there was a total of $(t_1 + t_2)$ sec between the emission of the light and its reception. That is to say, the total light-travel time from me to q was $(t_1 + t_2)$ sec. Since light travels at speed c ($= 3 \times 10^{10}$ cm/sec; here and hereafter we shall write c to denote this particular number of centimeters per second), the light, during

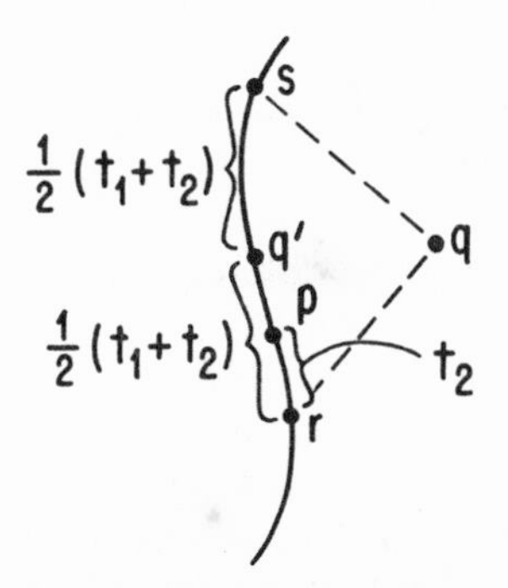

Fig. 39
Aristotelian interpretation of figure 38. An event q' is located halfway (in terms of elapsed time) between events r and s. This q' is interpreted as occurring simultaneously with q. Now, an apparent elapsed time and spatial distance between p and q can be calculated.

this $(t_1 + t_2)$ sec, had time to go a distance $c(t_1 + t_2)$ cm. But the light had to make a round trip, from me to q and back to me. So, the distance of q from me must have been $1/2\ c(t_1 + t_2)$ cm (see fig. 39). Finally, since p occurred right here by me, the spatial distance from p to q must be $1/2\ c(t_1 + t_2)$ cm." (Note: We are not committing ourselves at this point as to whether any of this is correct in any sense. We are just trying to get an intuitive idea for what the numbers t_1 and t_2 mean.) We now ask him a second question: "What was the elapsed time between p and q?" He might answer as follows. "Well, the light left at r, reached q, and then returned to me at s. So, since the light just went out and back, q must have occurred halfway between r and s (corresponding to the point q' in the diagram, although, of course, this person would not draw a space-time diagram). Since the elapsed time between r and s is $(t_1 + t_2)$ sec, the event q must have occurred just $1/2(t_1 + t_2)$ sec after r, for this is just half the elapsed time between r and s. On the other hand, p occurred just t_2 sec after r, for I measured this. Since q (for us, q') occurred $1/2\ (t_1 + t_2)$ sec after r, and since p occurred t_2 sec after r, q must have occurred $1/2(t_1 + t_2) - t_2$ sec after p. But $1/2(t_1 + t_2) - t_2 = 1/2\, t_1 + 1/2\, t_2 - t_2 = 1/2\, t_1 - 1/2\, t_2 = 1/2(t_1 - t_2)$. Thus, q occurred $1/2(t_1 - t_2)$ sec after p. This is the elapsed time between the events."

According to the paragraph above, one may, in a rough physical sense, think of $1/2\, c(t_1 + t_2)$ as the "spatial distance between p and q" (if you like, according to this clock), and of $1/2(t_1 - t_2)$ as the "elapsed time." Since the former involves the sum of t_1 and t_2, and the latter the difference, these two notions encompass all the information contained in t_1 and t_2. This, then, is our physical interpretation of these numbers.

Things have taken a remarkable and, from a certain standpoint, rather pleasant turn. We began with the Aristotelian and Galilean views, in which we were rather sloppy about the details of our geometrical constructions. We decided to be more careful. We allowed ourselves only two instruments and a few properties. As a result we obtain, as the relationship between events p and q, two numbers, t_1 and t_2. From these numbers we can reconstruct those notions which make sense in the Aristotelian view, the spatial distance $1/2\, c(t_1 + t_2)$ and the elapsed time $1/2(t_1 - t_2)$. In the Galilean view, on the other hand, only the latter makes sense. In short, we seem to be back where we started. In fact, however, we have made significant progress in two different directions. First, although we seem to obtain precisely the same notions we have always had, we now obtain these notions in a simpler and far clearer way. We understand what we are using, what we are doing, and what we are assuming. Second, and perhaps just as important, we have reexpressed these notions in a different way—in terms of the two numbers t_1 and t_2 with explicit physical significance.

All of this suggests that what we should now try to do is the following. We should regard the numbers t_1 and t_2 as carrying some information which refers to the observer (in particular, the choice of clock world-line) and some which is intrinsic to space-time. Indeed, in the Aristotelian view both t_1 and t_2 make sense, while in the Galilean only their difference makes sense.

What we want, then, is some other combination of t_1 and t_2 which is to make sense in our new view. We may be guided in this choice first by our experiences with the Aristotelian and Galilean views and second by the observations discussed in chapter 4.

Before we begin on this more important issue, we dispose of a preliminary problem. It is immediately clear that both t_1 and t_2 cannot be "intrinsic to space-time," by the following argument. Fix the events p, q, r, and s, and consider two clocks, A and B, each of which experiences events r, p, and s (fig. 40).

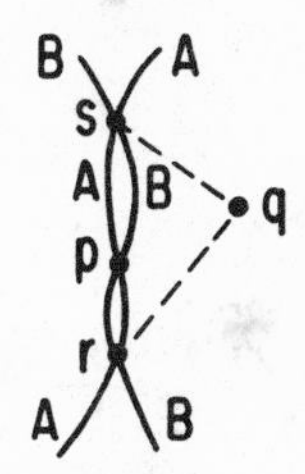

Fig. 40
Two clocks, A and B, both carrying out the measurement of figure 38. In this experiment, the clocks are so adjusted that their events p, q, r, and s are the same.

Let, according to clock A, the times in our construction be t_1 and t_2, and let those corresponding times according to clock B be t_1' and t_2'. It will certainly not be true in general that $t_1 = t_1'$ and $t_2 = t_2'$, for these two equalities would express the relationship between clocks which, at figure 34, we decided to reject. Thus, one clock assigns numbers t_1 and t_2 to events p and q, while another assigns t_1' and t_2' to the same events, and these times can be completely different. It appears almost as though our times are completely arbitrary. We shall get around this difficulty in the following way. Suppose one wanted to know the distance between two points on the earth, and one decided to determine this distance by asking someone to walk from one point to the other, keeping track of how far he walked. Such a procedure would be unsatisfactory, because the walker could take a very circuitous route between the points, inflating the distance. It would not even do to choose the points very close together, for even this choice would not prevent the walker from inflating the distance. The reason why even this last device fails, however, is this. We have worked ourselves into the situation in which we first commit ourselves as to the points, and then the walker decides what route he is to take. Suppose that we insisted that the decisions be made in the opposite order. Let us require first that the walker commit himself to his route, and only then will we commit ourselves as to how close together the points are to be. We can, in this "idealized limit of points close together, with 'how close together' only decided after the commitment of the walker" use this technique to obtain a reasonable measure of the distance between the points. If the walker's route, for example, is that of figure 41 (this figure in

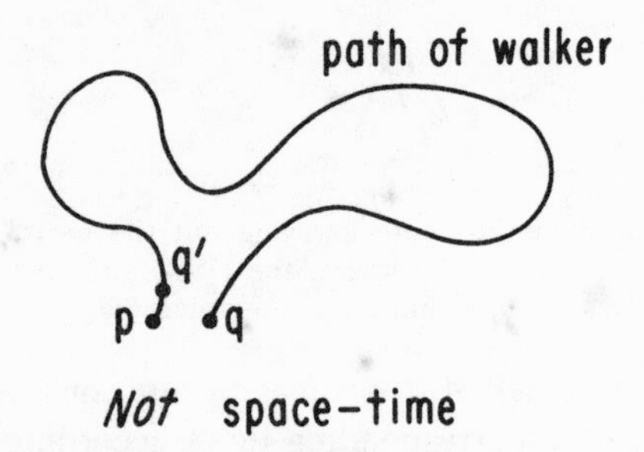

Fig. 41
A path of someone traveling from point p to point q. This is a diagram in space, *not* in space-time.

space, not space-time), then we might require that p and q be 1 mm apart, and in this case we would obtain a reasonable answer (for q could not then be where shown in the figure, but would have to be, say, at q'). (Another way to say the same thing: Ask the walker to commit himself to his turning radius when he walks [say, 100 cm]. This commitment having been made, we then choose how close together the points are to be [say, in this example, 1 cm].)

We shall avoid the difficulty illustrated in figure 40 in the same way. That is to say, we shall regard our construction (for

t_1 and t_2) as being carried out only in the limit when the events p and q are "nearby." As was the case with points on the earth, this statement by itself will not eliminate the problem. We therefore choose to interpret "nearby" in the same sense as we did for points on the earth, as follows: The event p and the clock world-line are to be chosen first, and only thereafter do we have to say how "nearby" q must be to p. This convention, then, essentially prevents the clock from "wiggling in space-time between r and p, or between p and s, distorting in one way or another the measured elapsed times between these two events." That is, with this convention certain details of the motion of the clock (namely, that illustrated in fig. 40) are not relevant to the resulting values of t_1 and t_2.

Remember that we are still in the process of trying to make the important decision, namely the guess, based on our experiences with earlier views, on our discussion of observations, and on any other evidence or hints we can obtain, of what combination of t_1 and t_2 is to be taken as intrinsic to space-time.

One further hint is available to guide us in making this guess. It is based on the observation that there are in fact five separate, distinct cases for our construction, depending on the relative locations of p and q (that is, on the order of r, p, and s along the clock world-line, since "location of q" is here translated, using the light-pulses, to events r and s on the clock world-line). These five cases are as follows.

The first case arises when the order of r, p, and s on the clock world-line is not that illustrated in figure 38 but rather that of figure 42, *A*. In this case, t_1 (the elapsed time between p and s) is positive, while t_2 (the elapsed time between r and p) is negative, for now p occurs before r.

The second case (fig. 42, *B*) arises when p and r coincide (that is, when the light-pulse from the clock to event q must actually leave the clock at event p itself). In this case, t_1 is positive, while t_2 (elapsed time from r to p) is zero, since r and p coincide.

The third case is that of figure 38. Event p occurs between r and s on the clock world-line, and both t_1 and t_2 are positive.

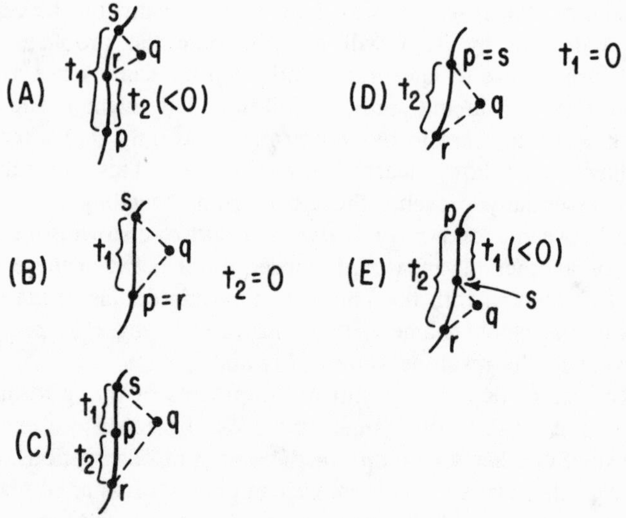

Fig. 42
Five cases of figure 38.

The fourth case (fig. 42, *D*) arises when events p and s coincide (that is, when the light returning from event q actually meets the clock at event p itself). In this case t_2 (elapsed time from r to p) is positive, while t_1 (elapsed time from p to s) is zero, since p and s coincide.

The fifth case (fig. 42, *E*) arises when both r and s occur before p. In this case t_2 is positive, while t_1 (elapsed time from p to s) is negative.

The five cases above have been distinguished in terms of the ordering of r, p, and s along the clock world-line, and in terms of the signs (and "zero-ness") of t_1 and t_2. In what sense does their existence provide us a hint in selecting our combination of t_1 and t_2? The point is that the question of which case one has depends only on the two events p and q, and not on the other details of the construction—in particular, not on the choice of clock world-line. The reasons are as follows.

In the first case (fig. 43, *A*), r occurs after p on the clock world-line. That is to say, a clock passes from p to r, and then light from r to q. Thus, light sent in all directions from p will encompass a region of space including q. That is to say, event q will lie inside the light-cone of p.

In the second case (fig. 43, B), r and p coincide, that is, a light-pulse sent from p reaches q. In this case, then, q lies on the light-cone of p.

In the third case (fig. 43, C), light, to reach q, must be emitted from the clock before p, while light emitted from q reaches the clock after p. This case thus corresponds to that in which p is neither on nor within the light-cone of q, and q is neither on nor within the light-cone of p.

In the fourth case (fig. 43, D), light from q reaches p. Thus, this case corresponds to that in which p lies on the light-cone of q.

In the fifth case (fig. 43, E), light from q reaches the clock before p. Thus, the event p will lie inside the light-cone of q.

Now the point of the above characterizations of the five cases is that they involve only the qualitative features of space-time:

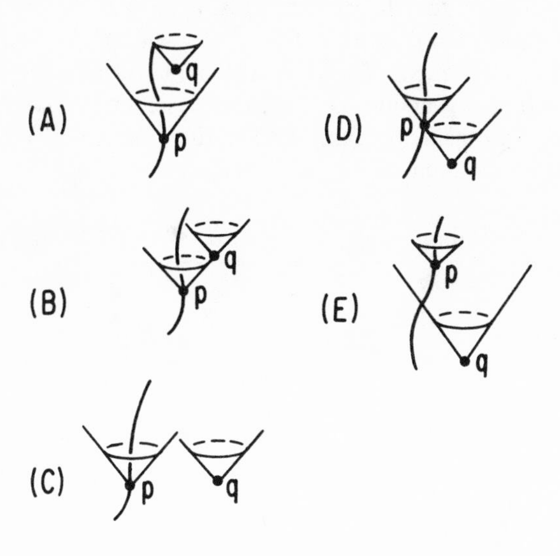

Fig. 43
The five cases of figure 42, now expressed in terms of light-cones.

light-cones, locations of events within, on, or outside those light-cones. These qualitative features do not refer directly to what the clock is doing at all. In other words, p and q themselves (and not the other details of the construction) already determine the particular case one is in. In short, given events p and q, the question of which of the five cases actually obtains is definitely "intrinsic to space-time." The five cases are conventionally referred to by the following terms. In the first and fifth cases (in which one of p and q lies within the light-cone of the other), p and q are said to be timelike related. In the second and fourth cases (in which one of p or q lies on the light-cone of the other), p and q are said to be lightlike related (or null related). In the third case (in which neither of p or q lies on or inside the light-cone of the other), p and q are said to be spacelike related. Furthermore, the first and fifth cases (in both of which p and q are said to be timelike related) are distinguished as follows. In the first case, one says that q is to the future of p, and in the fifth, that q is to the past of p. Similarly, the second and fourth cases are distinguished: In the second case, one says that q is to the future of p, and in the fourth that q is to the past of p. (There is neither reason nor necessity to make further distinctions in the third case, since neither event is either on or inside the light-cone of the other.) We summarize all these cases, then, in the table below.

Case	Name	t_1, t_2	Light-Cones
First	Timelike q future	$t_1 > 0 \quad t_2 < 0$	q in light-cone p
Second	Lightlike q future	$t_1 > 0 \quad t_2 = 0$	q on light-cone p
Third	Spacelike	$t_1 > 0 \quad t_2 > 0$	None
Fourth	Lightlike q past	$t_1 = 0 \quad t_2 > 0$	p on light-cone q
Fifth	Timelike q past	$t_1 < 0 \quad t_2 > 0$	p in light-cone q

The situation, then, is as follows. We have distinguished and named five cases for the relationship between p and q. The question of which case one has (given p and q) is "intrinsic to space-time," since it depends only on the light-cones of points. On the other hand, these cases can also be described in terms of the signs of the numbers t_1 and t_2. We are looking for some combination of t_1 and t_2 that we may regard as intrinsic to space-time. What we know, from the discussion above, is that the signs are certainly intrinsic. The hint, then, is that one might expect that the signs of t_1 and t_2 should somehow be prominently displayed in whatever combination we choose, that these signs should be "appropriately reflected" in our combination, or "respected by" our combination. (I do not know how to put this matter more precisely. I personally do not believe that one can—or even that one should try to—"deduce" the proper combination. It is a matter of making an informed guess. Presumably, not everything about t_1 and t_2 will be intrinsic. One might at least hope, however, that our ultimate choice of combination will reflect everything which is intrinsic about t_1 and t_2. Since we have found something which is intrinsic—the signs —one might hope that at least this will be reflected in our combination.)

All this is leading up to our final guess, which is this: It is the product, t_1t_2, which is to be intrinsic. (Note how "respectful" products are of the signs of the factors.) This product, t_1t_2, is called the interval between p and q. The supposition, then, is that this interval is intrinsic to space-time; it is independent of all the details of the construction (subject, of course, to our earlier convention about the events being "nearby"). In other words, no matter what clock world-line is chosen, no matter whether it passes through p or through q, and so on, we shall, it is now supposed, always obtain the same interval, t_1t_2, between p and q.

In real life, one would regard this supposition—that it is the interval, so defined, which is intrinsic—as a sort of tentative working hypothesis. One would then proceed to work out various consequences, implications, and predictions of this "working hypothesis," comparing them, whenever possible, with our world. If these consequences are found to generally agree with what we see about us, and if in particular they do not in any case violently disagree with what our world is like, then one's faith in this working hypothesis would, over time, become stronger. As more and more observations can be understood in terms of the working hypothesis, one's faith becomes still stronger. If it should happen at some point that disagreements arise between the hypothesis and what we see in the world, one would normally first try to resolve them within the hypothesis by some other explanation. If not found possible, one would try a new hypothesis, and begin over. If a long succession of such working hypotheses were tried, and none seemed to work, one might abandon this whole approach (in terms of space-time), and look for some entirely different approach (of which, unfortunately, I cannot offhand think of any examples). If even a variety of "entirely different approaches" did not seem to work, one would, in practice, probably abandon the subject completely, or at least would considerably reduce the time one spends in it. (One likes to spend one's time where one is likely to make some progress.) The point is that in things of this kind nothing is ever "proven." Everything is essentially a working hypothesis in one form or another. Some working hypotheses are rather tentative, and in some one has a rather firm conviction. But there are always caveats, the sands are constantly shifting, one never finds absolutely solid ground on which to stand. We, unfortunately, do not have time to explore these subtleties of "real life." We jump to what is the accepted "guess" and explore its consequences.

In a certain sense, we have now completed the statement of the general theory of relativity. (Actually, there is one more working hypothesis which must be made later, but it is of rather minor importance compared with what we have done so far.) What remains, essentially, is the working out of consequences of what we have, and the comparison of those consequences with what we see about us in the world. It is this program that will occupy us for some time.

Before embarking on this program, it may be worthwhile to review how we have gotten to where we are. We have proceeded in essentially two steps. The first step (in my view, the more important) was the acceptance of the general framework. One should introduce events, assemble them into space-time, and describe goings-on in the physical world in terms of collections of events and relationships between events. One should not carry over directly all one's intuitive ideas about space and time to structure on space-time. Rather, one should introduce and characterize in terms of space-time the measuring instruments one wishes to use, and require that whatever structure one ultimately obtains on space-time be carried into this framework by the instruments themselves. One should search for some type of geometrical structure on space-time which will serve to reflect the ideas we all have about space and time. The second step is perhaps just the technical completion of the first. One decides that an appropriate choice of instruments is light-pulses and clocks, and what the description and properties of these instruments are to be. One decides that one's relationships between events will be obtained, using these instruments, via the construction illustrated in figure 38. Finally, one decides that the intrinsic relationship—the geometrical structure to be induced on space-time—will be the interval. (Of course, one is not expected at this point to "believe" that the interval is a good choice. This belief, it is hoped, will come later as we investigate consequences.)

We have finally arrived at what is taken as the "intrinsic relationship between events" in the relativistic view. The relationship is a certain number (of seconds squared), the interval, associated with any two nearby events; it is intrinsic in the sense that, although it is determined operationally in terms of measurements made by a certain individual, this number is asserted to be the same, given the two events, no matter who measures it. It is now our purpose to begin the program of interpreting what this interval means physically.

We will postpone our discussion of physical interpretations for a moment to insert a few remarks on a general theme, to which we shall return on several occasions. This theme is the broad picture of space-time that one adopts in the relativistic view. Let us suppose that one wished to describe the geometry of the surface of the earth. As a first step, one might construct a catalog of all points on the earth (that is, "Chicago," "this point in the middle of the Pacific," and so on). The mere listing of all these points does not, however, exhaust the relevant geometrical information about the surface; one wishes in addition to specify some distances. One might therefore proceed as follows. One prepares a second catalog, each entry of which gives the following information: one point from the first catalog, a second point from the first catalog, and the distance, along the surface of the earth, between those points. For example, one entry might read "Chicago, New York, 900 miles." This second catalog would include the distances between all pairs of points listed in the first catalog. Such an approach to the geometry of the surface of the earth would, I claim, not be wholly satisfactory. The problem is that one must specify, not only the distance between the two points, but also some indication of the route along which that distance is measured (for example, "along route 80," "as flown by United Airlines"). In short, unless one can think of some general convention for how the distances are to be measured, our second catalog will contain an enormous number of footnotes.

There is, however, a much simpler way of describing the geometry of the surface of the earth. We keep the first catalog. The second, however, is modified as follows: We only make entries corresponding to distances between nearby points on the earth (for example, "Chicago, Downer's Grove, 24 miles"). As we have already remarked, such distances are unambiguous for nearby points (provided that we adopt the convention described earlier: commitments as to instruments are made prior to commitments as to "how nearby"). Thus, the second catalog is now much smaller (for, since we now have only to include pairs of nearby points, there are many fewer entries), and furthermore has no footnotes (for our distances are now unambiguous). We may take these two catalogs, then, as describing the geometry of the surface of the earth. In what sense does this more limited information cover everything one would want to know about the geometry? In fact, it has as much information as was in the two original catalogs. Suppose that we wanted to determine the distance from Chicago to New York. Our first question would be "Along what route?" Let that question be answered by the specification of some route. We would then find a sequence of successively nearby points, from Chicago to New York, along that route. (This sequence might, for example, be "Chicago, Hammond, Gary . . . Newark, New York.") One then looks up, in one's catalog, the distance from Chicago to Hammond, that from Hammond to Gary . . . that from Newark to New York. (These entries should all appear in our second catalog, since they refer only to nearby points.) Adding these distances, one obtains the distance from Chicago to New York along this particular route. Thus longer distances can be obtained by adding the distances between nearby points.

The relativistic view of space-time is rather similar to the above description of the geometry of the earth. One thinks of "space-time" as consisting of two catalogs. The first lists all possible events (from the past, the present, and the future; from everywhere). The second catalog lists all pairs of nearby events, and, for each pair, the interval between them. In this sense, space-time is a "geometrical entity." The fact that our catalog only lists intervals between pairs of nearby events does not preclude our speaking ultimately of relations between nonnearby events. Such relationships, however, will require the specification of something (such as a path in space-time, that is, a world-line) connecting the events. Using such a connecting entity, relationships between nonnearby events can be expressed in terms of relationships between a succession of nearby events (the latter relationships being, of course, given in our second catalog). Again, the above is intended as merely a general framework, to be fleshed out as the occasion arises.

We return now to the physical interpretation of the interval. Our first task is this. We know the expression for the interval between two fixed events in terms of the two times, t_1 and t_2, recorded by the clock (namely, $t_1 t_2$). Furthermore, we know the formulas, in terms of t_1 and t_2, for the "apparent spatial distance between the two events" (namely, $\triangle x = c/2\ (t_1 + t_2)$) and the "apparent elapsed time between the two events" (namely, $\triangle t = 1/2\ (t_1 + t_2)$). Recall that *apparent* here means "as would be the judgment of someone who remains with the clock, who knows about speeds and distances but does not know anything of relativity or space-time, and who believes that light travels at speed c." Our task is to express the interval in terms of these apparent distances and apparent times. We solve this algebraic problem as follows. We first divide both sides of the formula for the apparent distance by c, to obtain

$$\triangle x/c = \tfrac{1}{2}(t_1 + t_2).$$

We next just copy the formula for the apparent time:

$$\triangle t = \tfrac{1}{2}(t_1 - t_2).$$

Adding these two equations, we obtain

$$\triangle x/c + \triangle t = t_1.$$

Subtracting,

$$\triangle x/c - \triangle t = t_2.$$

What are we doing? We are regarding our two equations—for the apparent distance and the apparent time—as two linear equations for the two "unknowns," t_1 and t_2 (taking $\triangle x$, $\triangle t$, $1/2$, and c as "knowns"). We are solving this set of two linear equations in two unknowns for t_1 and t_2. The solution is the two formulas immediately above. Thus, so far we have expressions for t_1 and t_2 in terms of $\triangle x$ and $\triangle t$. But "interval $= t_1 t_2$," and so we can just substitute (for t_1 and t_2 in terms of $\triangle x$ and $\triangle t$) to find the interval in terms of $\triangle x$ and $\triangle t$. The calculation is as follows:

$$\text{Interval} = t_1 t_2 = (\triangle x/c + \triangle t)\ (\triangle x/c - \triangle t)$$
$$= [(\triangle x)^2/c^2] - (\triangle t)^2.$$

The first equality is the expression for the interval, the second substitution, and the third expansion.

This is our final result. It means that our nonrelativist tagging along with the clock would say "Well, this 'interval' that you people seem so interested in is computed as follows. I take the square of the (apparent) spatial distance between the events and divide by the square of $c = 3 \times 10^{10}$ cm/sec. Then, I subtract the square of the (apparent) elapsed time between the events." We, of course, know what this nonrelativist does not know, namely that different individuals (that is, with different clock world-lines, and so forth) making similar measurements on these same two events will come up with different "apparent spatial distances" and "apparent elapsed times." On these things they will disagree completely. The only thing they will agree on is this strange combination, $[(\Delta x)^2/c^2] - (\Delta t)^2$, the interval.

We may represent the situation graphically (fig. 44). Fix once and for all the two events, p and q. Let someone with a clock and flashlight measure t_1 and t_2, and from these compute an apparent spatial distance and an apparent elapsed time. The result will be two numbers. We may represent these two numbers on a piece of graph paper with one axis labeled "apparent

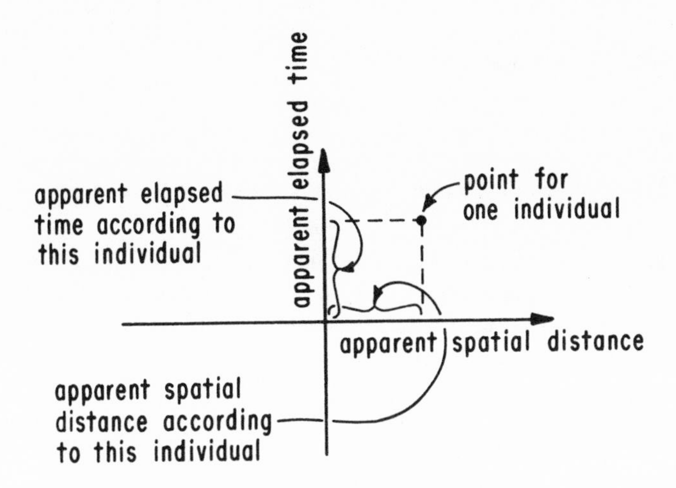

Fig. 44
Fix two events. Each determination by an observer of an apparent spatial distance and an apparent elapsed time between the fixed events yields a point on the graph.

spatial distance" and the other "apparent elapsed time." This individual's numbers define a point on the paper. A second individual, using the same two events, will obtain different apparent things, and so will obtain some other point on the graph paper. If many individuals are permitted to perform this experiment (always on the same two events), we shall obtain a number of points on the graph paper.

Question: Where will all these points lie? Different views of space-time (fig. 45) lead to different answers. The Aristotelian

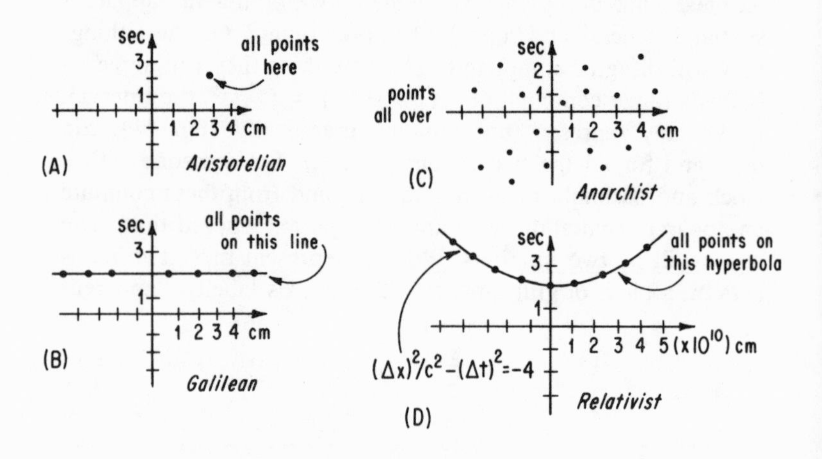

Fig. 45
Four possibilities for the result of the plot of figure 44.

(fig. 45, *A*) feels that spatial distance and elapsed time are both "intrinsic to space-time," that is, the same for everyone. He, therefore, would feel that all the points should be the same, that everyone would determine the same distance and time between the events. The Galilean (fig. 45, *B*), more sophisticated, realizes that "spatial distance" depends on who does the looking. He therefore realizes that not all points must have the same "apparent spatial distance" coordinate. However, the Galilean does feel that "apparent elapsed time" is the same for everyone, that everyone's point in the diagram would have to have the same "apparent elapsed time" coordinate. In short, the Galilean would claim that all the points on the diagram must lie along a horizontal line, as shown. Someone who believes that Nature is capricious, that this whole program is a waste of time, that we shall never find any patterns in anything, might claim that the points would be scattered all over the diagram (fig. 45, *C*). (Of course, if experiment showed this to be the case, we would have to go back to the drawing board.) What would we, as relativists, claim that the diagram will look like? We would claim that everyone will compute the same interval, that is, that all the points will lie on a single curve whose equation is of the form $[(\triangle x)^2/c^2] - (\triangle t)^2 = const$, where *const* denotes some number (fig. 45, *D*).

One important difference between figure 45, *D* and figures 45, *A*, *B*, and *C* should be noted. Whereas in every graph the unit of time is chosen to be the second, different units for distance are used. The first three use the unit centimeters, and the last the unit "10^{10} cm." Why have we done this? As an example, let $\triangle x$ be 3 cm and $\triangle t$ 2 sec. Then the interval would be as follows: interval $= [(\triangle x)^2/c^2] - (\triangle t)^2 = [(3)^2/(3 \times 10^{10})^2] - (2)^2 = 10^{-20} - 4 = -3.99999999999999999999$. The answer is practically "-4." The reason is that c is so large (3×10^{10}) that, if $\triangle x$ is some reasonable number (such as "3" in our example), then $(\triangle x)^2/c^2$ will be very small indeed $(10^{-20} = 0.00000000000000000001$ in our example). This first term, $(\triangle x)^2/c^2$, will thus make practically no contribution to the interval; the total interval will be essentially just $-(\triangle t)^2$. The point, then, is that the first term in the formula for the interval just won't make much difference if we plot the relativistic graph with the units "seconds" and "centimeters." Indeed, we may plot the graph in the relativistic case with these units, and the result is figure 46. This graph results from figure 45 in that it has been "stretched horizontally by a factor of 3×10^{10}." All this stretching pulls out the hyperbola in the earlier graph, making it appear as a nearly horizontal line. The reason, then, that we used a different unit for distance in the last graph was in order to make the hyperbolic shape visible.

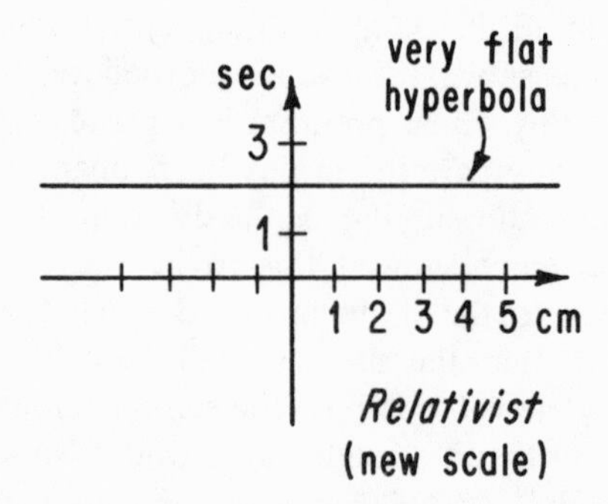

Fig. 46
Figure 45, *D*, replotted on the same scale as figure 45, *A-C*.

But this is the whole point. Using any "reasonable" units, the graph in the relativistic case looks almost identical to the graph in the Galilean case. That is to say, what is "intrinsic to space-time" in the relativistic view differs, under normal circumstances, very little from what is intrinsic in the Galilean view. In still other words, the geometry and physics of space-time in the relativistic view differ only slightly from the geometry and physics in the Galilean view. We may now understand how it could happen that the Galilean view could come to be so popular. If one uses only crude instruments (as one certainly does in everyday life), then one has little hope of telling the difference between a straight horizontal line and a nearly straight hyperbola for the true "apparent spatial distance-apparent elapsed time" relationship. Suppose for a moment that the relativistic view is the correct one, and that the actual graph is our hyperbola. Then, for everyday purposes, an approximation by a straight horizontal line is perfectly adequate. It is in terms of this approximation to the relativistic view—the Galilean view—that we are used to thinking. Indeed, it would be shocking if something like this did not happen. One can criticize the Galilean view all one wants, but the fact remains that it is useful and rather accurate for everyday life. We now see that relativity, in some sense, "predicts" this usefulness.

As part of the motivation for the introduction of the interval, we introduced the distinction between events lightlike, timelike, and spacelike separated. As a further aspect of the physical interpretation of the interval, let us now ask our nonrelativist with the clock what he would have to say about this distinction. So, let us tell our friend that two events are lightlike related, asking for his interpretation; and, in answer to his query as to what "lightlike" means, let us answer that it means that the interval is zero. He might then reply as follows. "Well, this interval of yours is given by the formula $[(\triangle x)^2/c^2] - (\triangle t)^2$. You are telling me that it is zero, so I have $[(\triangle x)^2/c^2] - (\triangle t)^2 = 0$, or $(\triangle x)^2/c^2 = (\triangle t)^2$, or (taking the square root) $(\triangle x)/c = \triangle t$, or (multiplying by c) $\triangle x = c\triangle t$. It is this relationship I am to interpret. Now, I know that in a time $\triangle t$ light will manage to travel a distance given by $c\triangle t$. (Note that he does not have the same qualms that we do in talking about distances and times.) But what I have is just $\triangle x = c\triangle t$. So, my interpretation is as follows: During the elapsed time between the two events, light has just enough time to travel a distance equal to the spatial distance between the two events. In other words, my interpretation is that light is just able to make it from one event to the other." Now, the remarkable thing about his reply is that it is exactly the same as our earlier physical interpretation of "lightlike related" (for example, in fig. 42, *B* and 43, *B*). We also interpret "lightlike related" to mean that light just makes it from one event to the other (that is, that one event lies on the light-cone of the other). In short, we have now come full circle. We began with the notion "light just makes it from one event to the other," then expressed this notion in terms of t_1 and t_2, then expressed it in terms of the interval. We then presented this expression (in terms of the interval) to our nonrelativist friend, and he ends up interpreting it as "light just makes it from one event to the other." There is no deep point to be made here; it is just that these are all different ways of saying the same thing. Let us now do the same thing for "timelike." In this case, the nonrelativist might reply as follows. "Well, you are now telling me that $[(\triangle x)^2/c^2] - (\triangle t)^2 < 0$, or $(\triangle x)^2/c^2 < (\triangle t)^2$, or $(\triangle x)/c < \triangle t$, or $\triangle x < c\triangle t$. Since, once again, light, in time $\triangle t$ can go a distance $c\triangle t$, my interpretation is this: During the elapsed time between the two events, light has enough time to travel farther than the spatial distance between the two events. In other words, light, emitted from one event, will actually get out beyond the other event before it occurs." But this, again, is just our physical interpretation of the relation "timelike" (for example, in fig. 43, *A*). In figure 47 we give a little more detail, including the worldline of someone who is waiting around for the occurrence of event *q*. This person experiences the light from *p* at event *u*, an event he experiences before *q*. That is, this person waiting around would feel that *q* occurred too late to be on the lightcone of *p*, that is, that the light from *p* had already gone by before *q* occurred. Finally, we do the same thing for "spacelike," with the following reply. "Well, now we have $\triangle x > c\triangle t$, and so my interpretation is this: During the elapsed time between the two events, light had enough time to travel a distance

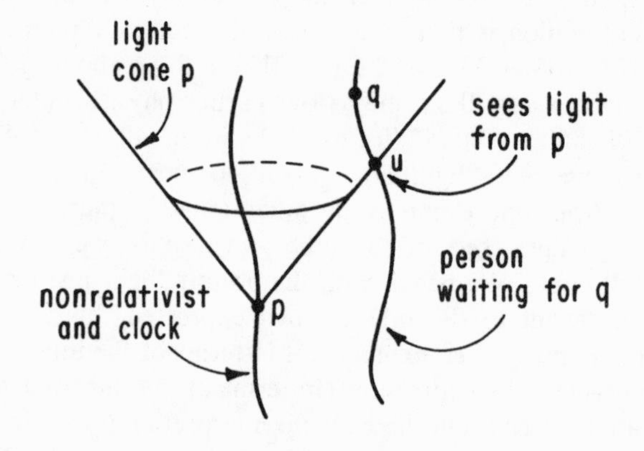

Fig. 47
Timelike-related events *p* and *q*, as interpreted by a nonrelativist.

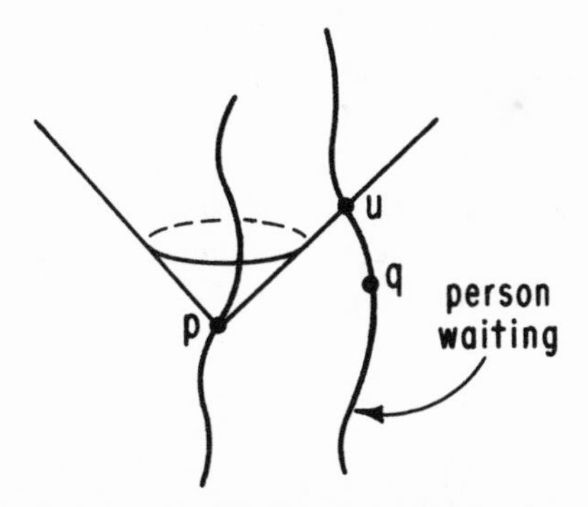

Fig. 48
Spacelike-related events p and q, as interpreted by a nonrelativist.

less than the spatial distance between the two events. In other words, light, emitted from one event, will never make it to the other event, for the other will occur too soon." Again, this is the same as our physical interpretation (for example, fig. 42, *C*). Again, we draw the diagram in more detail, including the world-line of someone who is waiting around for the occurrence of event q (fig. 48). This person first experiences event q, and only thereafter does he experience (at event u) the light coming to him from event p. This person, then, would feel that q occurred too early, that is, before the light from p had time to reach him.

We are in the process of trying to interpret the interval physically, to give some indication of what thoughts might come to mind if it were announced that the interval between two events is such-and-such a number of seconds squared. The first interpretation involved the relation between the interval and the representation on a graph of the apparent spatial distance and elapsed time (according to a nonrelativist with the clock) between the events. The second involved the interpretation of the terms *timelike*, *lightlike*, and *spacelike* by a nonrelativist. ("Interpretation" normally means "by a nonrelativist," because the former has connotations of "interpretation in everyday terms," while our everyday life seems nonrelativistic.) Finally, we now give an alternative interpretation of *timelike* and *spacelike*.

Let two nearby events p and q be timelike related, say, with q within the light-cone of p. Then, since q is within the light-cone of p, one might expect to be able to arrange matters so that some clock directly experiences both of the events p and q, as in figure 49, A. Let such a clock be introduced, and let us

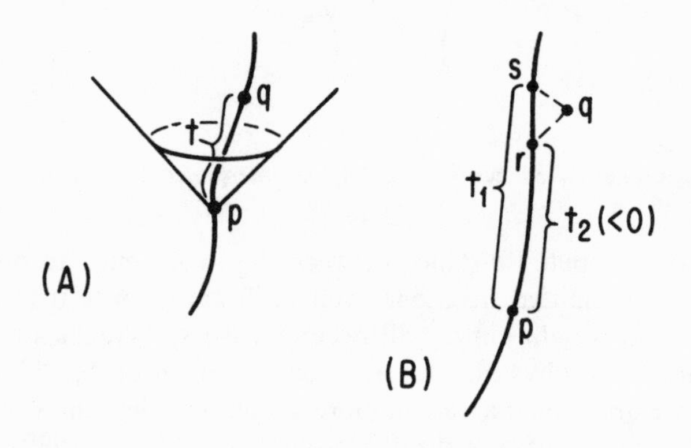

Fig. 49
A is the limiting case of B as q approaches the world-line of the clock. The interval between p and q in A, therefore, is just $-t^2$.

denote by t the elapsed time, according to that clock, between events p and q. Our goal is to express the interval between p and q in terms of this measured time t (that is, some number of seconds). To accomplish this, we consider the situation in figure 49, B. This is, of course, merely a special case of our basic construction (fig. 38), when q happens to be very close to the world-line of the clock. In this special case, the points r and s on the world-line of the clock are very close together and also close to q. It should now be clear that figure 49, A is just the limit of figure 49, B as q approaches the world-line of the clock. That is, in figure 49, A, q, r, and s all coincide. Thus, for that diagram, we have $t_1 = t$ (elapsed time from p to $s = q$) and $t_2 = -t$ (elapsed time from $r = q$ to p). Hence, the interval $= t_1 t_2 = -t^2$. This is the desired formula. In words, it says that if you happen to have a clock on you, and if you happen to experience two nearby events (on your world-line), you can easily compute the interval between them by taking minus the square of the elapsed time you measure between those two events. Suppose, alternatively, that you were told that the interval between two events is, say, -100 sec^2. This could be interpreted as follows. Since, first, the interval is negative, we know that the events are timelike related. Suppose that one arranged to carry a clock, and to directly experience both events. Then the elapsed time between those two events, according to that clock, would be 10 sec (for $-(10)^2 = -100$).

We next interpret in a similar way "spacelike related." Let events p and q be spacelike related. Now one could imagine sending many different clocks through event p. Some might observe a t_1 value greater than t_2, while others might register t_2 greater than t_1. Let some clock be so arranged that $t_1 = t_2$, say that in figure 50. Again, let us let our nonrelativist travel with this particular clock, and let us ask him what he has to say about the relationship between the two events. He, as always, would want to compute the elapsed time and the spatial separation between the two events. For the elapsed time, he would compute $\triangle t = 1/2(t_1 - t_2)$. But this is zero, since we have arranged matters so that $t_1 = t_2$. Our nonrelativist, then, would say that the two events occurred "at the same time." He would obtain the spatial separation (according to him) by $\triangle x = c/2$ $(t_1 + t_2)$, some number of centimeters. So, our nonrelativist's interpretation so far is that "the two events occurred at the same time, but separated by a spatial distance $\triangle x$." We now compute the interval. We have interval $= (\triangle x)^2/c^2 - (\triangle t)^2$ $= (\triangle x)^2/c^2$, where we used $\triangle t = 0$ in the last step. This is the desired formula. It says that the interval between the two events is just the square of the spatial distance between them (as determined by a nonrelativist who feels that they occurred "at the same time") divided by the square of $c = 3 \times 10^{10}$.

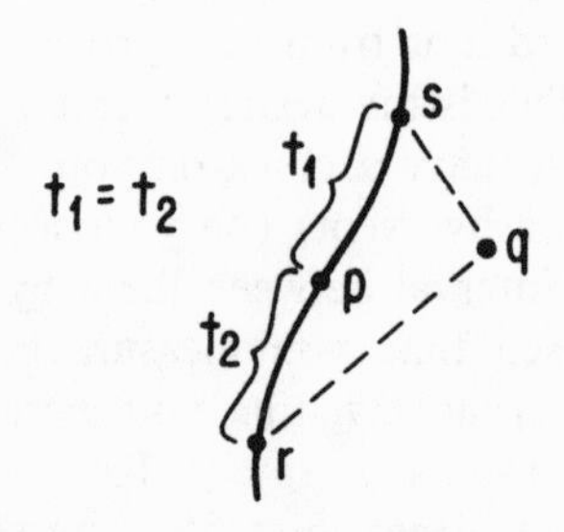

Fig. 50
A clock is so arranged that it experiences event p, and it has $t_1 = t_2$. Then the interval between p and q, according to this clock, is the square of this common time.

Suppose, as an example, that you were told that the interval between two events is, say, 10^{-18} sec². This might be interpreted as follows. Since the interval is positive, the events must be spacelike separated. Suppose, then, that one arranged to carry a clock, and that one arranged himself so that he felt that the two events (one directly experienced and one not) occurred simultaneously. By "felt . . . occurred simultaneously" we mean, of course, that, on sending light to the event off one's world-line and receiving the return signal, the times t_1 and t_2 were equal, as in figure 50. Then from t_1 and t_2 one could compute an apparent distance, $\triangle x = c/2\ (t_1 + t_2)$. Since we know that the interval is supposed to be 10^{-18} sec², we must have $(\triangle x)^2/c^2 = 10^{-18}$, or $(\triangle x)^2 = 10^{-18}\ c^2$, or $(\triangle x)^2 = 10^{-18}\ (3 \times 10^{10})^2$, or $(\triangle x)^2 = 10^{-18}\ (9 \times 10^{20})$, or $(\triangle x)^2 = 900$, or $\triangle x = 30$ cm. One, having arranged himself so that he regards the events as occurring simultaneously, would regard the events as having a spatial separation of 30 cm.

We see from the discussion above that the interval has two types of interpretations, depending on the relationship between the events. For the case of events timelike related, the interval is a measure of an apparent elapsed time between the events. (We mean *measure of* in the sense that the interval is not the elapsed time itself, but rather is given by the formula, interval

$= -$(apparent elapsed time)^2. By *apparent* we mean apparent to that observer who happens to see both events on his world-line.) For the case of events spacelike related, the interval is a measure of an apparent spatial distance between the events. (We now mean *measure of* in the sense that the interval is given by the formula, interval $=$ (apparent spatial distance)$^2/c^2$. By *apparent* we now mean apparent to that observer who happens to see the events as occurring simultaneously.) So, the interval is "sometimes a measure of an apparent elapsed time and sometimes a measure of an apparent spatial distance." More generally, it is a certain combination of these two. There seems to be no similar interpretation of the interval in the lightlike case. There, "two events lightlike separated can be joined by a single light pulse" will have to do.

One might have noticed that, in the examples above, the timelike interval we chose was reasonable-sized (although, of course, negative), namely, -100 sec^2, whereas the spacelike interval was very small, 10^{-18} sec^2. However, when everything finally got expressed (by our nonrelativist) in terms of normal units, seconds and centimeters, the numbers were quite reasonable, namely 10 sec and 30 cm. How does it happen that "reasonable spatial distances" (according to the nonrelativist) come out to very small spacelike intervals, while "reasonable elapsed times" (nonrelativist) come out to reasonable timelike intervals? The reason, of course, is that $c = 3 \times 10^{10}$ is so large in terms of normal units. These remarks in turn suggest that, in everyday terms, a spacelike interval is a rather delicate sort of thing compared with a timelike interval. We may make this point more dramatically as follows.

You are told, "You have two hands; each can be used, by a finger-snap, to mark an event. Mark two events which are timelike related." This is very easy to do. As soon as you hear the command, you snap, say, with your left hand. Then, a couple of hours later (or days later, or whenever you get around to it), you snap with your right hand. The resulting two events will, certainly, be timelike related. Suppose, however, that you are now told "Mark two events which are spacelike related." You know what you must do. You must arrange matters so that a pulse of light, emitted from your left hand as it snaps, is not able to get over to the right hand before it snaps, and, conversely, light from the snapping in the right hand is unable to reach the left before it snaps. Clearly, you cannot wait days, or even hours between the snap of one hand and the other, for "hours" gives light plenty of time to get from one hand to the other. In fact, you have to snap your fingers very nearly "at the same time." So, you stretch your arms out (to place as much burden of traveling on the light as possible), and try to snap your fingers at the same time on both hands. But, human error being what it is, one hand snaps, say, 1/100th of a second (we're nonrelativists now) before the other. Those two events were timelike related. In fact, you've got to snap your fingers within about a billionth of a second of each other. It's not at all easy; marking with your hands two events spacelike related could well be an Olympic event. The point, then, is that "practically any two events" are going to end up being timelike related. By the interpretation above, then, the interval can normally (that is, "practically always") be interpreted as simply minus the square of an apparent elapsed time between the events. The "$\triangle x$ term" in the formula for the interval just doesn't make much difference in everyday life. But the elapsed time is precisely what is intrinsic in the Galilean view. Again, we can now understand how it could happen that the Galilean view would come to be the dominant one in everyday life.

We have developed a certain attitude toward space-time. One thinks of space-time as consisting of two books, the first listing all possible events in the world and the second listing, for all possible pairs of nearby events, the interval between those events. These two books, so the attitude goes, are to contain all the spatial and temporal information there is to be had within our world. This information is to be decoded as follows. Physical phenomena are to be described in terms of collections of events (for example, world-lines for particles, world-surfaces for ropes, and so on). The intervals in the second book, then, provide information about relationships between events and so, since physical phenomena are described in terms of events, provide "spatial and temporal statements" about physical phenomena. It is only through the book of intervals, then, that we are permitted to say anything having a spatial or temporal content. This, at least, is our stance. We now, however, are presented with a marvelous opportunity to test it. We already have before us two examples of specific "physical phenomena," the light-pulses and clocks that we introduced earlier. So, we ask, can we describe the behavior of light-pulses and of clocks in terms of the interval? Were the answer no, we would presumably have to give up our attitude; fortunately, it is yes. We now see how this comes about.

We begin by stating in a little more detail what we are trying to do. Imagine someone who has spent his entire life in a closet, having virtually no opportunity to see what our world is like. This person has, however, been taught relativity. Those on the outside have, after an enormous expenditure of time and effort, managed to compile two books, one listing all events in the world, the other all intervals between pairs of nearby events. Now, the people outside actually know much more than is contained in the book. They, for example, can shoot light-pulses around, and find their world-lines; they can build clocks, and find their world-lines and the corresponding time-functions associated with those clocks. The situation we wish to consider, then, is the following. The two books, and only those two books, are passed in to the person in the closet. He is now instructed to try to figure out from just those books what the people outside already know; namely, what are the world-lines of light-pulses and what are the world-lines and times for clocks. The question, then, is whether or not there will be enough information in the books to allow him to figure this out.

We begin with the light-pulses, since they are easier. The person in the closet first makes the following definition: I call a world-line lightlike if any two nearby points along that world-line are lightlike related. It should be emphasized that this is a definition he may make, for it refers only to the information, namely the two books, that he has in his possession. A world-line for him is of course just a collection of events in the first book. For such a world-line, he can take all pairs of nearby events on that world-line, look each pair up in the second book, and see whether or not the interval is zero. If, for every such pair, the interval is zero, then he agrees to call that world-line lightlike. In short, he does not have to look at anything physical (such as actual light-pulses) to make this definition. But the person in the closet also knows relativity. He knows, among other things, the following fact: Given two lightlike related events, a single light-pulse can experience them both. That is to say, he knows, from his knowledge of relativity, that the actual world-lines of light-pulses will be his lightlike world-lines. Having thought this all through, he now announces to those outside an entire collection of world-lines (namely, those which satisfy his definition of lightlike), and claims those to be the world-lines of light-pulses. This, of course, is the right answer, as those on the outside proceed to verify. What has happened, then, is that our person in the closet has correctly "predicted" what the world-lines of light-pulses will be, knowing only what is in the interval. In this sense, information about the behavior of light is already carried by the interval.

We now do the same thing for clocks. Our person in the closet in this case makes the following definition: I call a world-line timelike if any two nearby points along that world-line are timelike related. (Again, this makes sense given only the books.) Since he knows relativity, he knows the following: Any two nearby points along the world-line of a clock are in fact timelike related. (See, for example, the discussion of figure 49.) In this way, then, he determines what the world-lines of clocks will be. He announces all these world-lines to those on the outside, and they proceed to verify his prediction. This, however, does not complete the story for the case of clocks. Although our person in the closet has correctly determined what the world-lines for clocks are to be (a determination of course subsequently verified in the world outside), he has not yet said anything about the elapsed times as measured by those clocks. Can he determine, from only his two books, what the time-readings of the clocks will be? He does so as follows. Take a particular timelike world-line, and fix two points, p and q, on that world-line. (This, of course, is something that our person

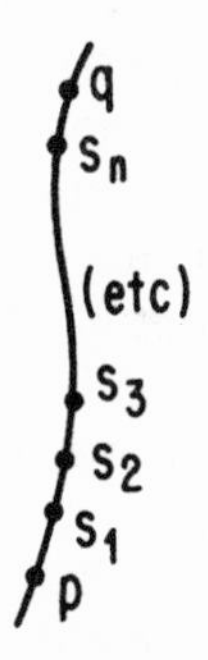

Fig. 51
The computation, using only the interval, of the elapsed time, along a certain world-line, between events p and q. One introduces a succession of closely spaced events along the world-line, computes from the interval the elapsed time between successive events, and adds.

in the closet can do.) He now wants to predict, from his books, what the measured elapsed time between p and q, along this world-line, will be when the experiment is actually performed in the world. He chooses a sequence of points, marked s_1, s_2 . . . s_n, along the world-line from p to q (fig. 51). He chooses these points so that p and s_1 are nearby, s_1 and s_2 are nearby, s_2 and s_3 are nearby, and so on until s_n and q are nearby. Next, he gets out his book with the intervals. He looks up "p and s_1" in that book, and finds the interval between those points. It will, of course, be negative, since p and s_1 must be timelike related (by definition of a timelike world-line). He then takes minus this interval, and takes the square root, to obtain a certain number of seconds. Next, he looks up s_1 and s_2, finds the interval, takes minus it, and takes the square root again, to obtain some other number of seconds. He continues in this way, finally taking the square root of minus the interval between s_n and q. The results of these calculations are various numbers of seconds. Finally, he adds all these numbers together (that is, the square root of minus the interval between p and s_1 plus the square root of minus the interval between s_1 and s_2, and so on, until . . . plus the square root of minus the interval between

s_n and q). The grand total is some number of seconds. But our person in the closet also knows relativity. He therefore knows, among other things, the discussion on page 105. He knows that the square root of minus the interval between p and s_1 is just the elapsed time between p and s_1 as measured by the clock in figure 51; that the square root of minus the interval between s_1 and s_2 is just the elapsed time between s_1 and s_2 as measured by the clock in figure 51; and so on. He knows, therefore, that the sum he has just done will be precisely the elapsed time between p and q as measured by this clock. He therefore announces this sum to the people on the outside: "The elapsed time between this event p and that event q as determined by the clock along this particular world-line will be so many (the number he calculated) seconds." Those on the outside proceed to check his claim, and of course find it correct. Our person in the closet, in short, is able to determine, from only his two books, what will be the physically measured elapsed time between two events for a given clock world-line. This is in addition to his being able to determine the clock world-lines themselves. In this sense, then, information about the behavior of clocks is already carried by the interval.

Our conclusion, then, is that from the interval one can determine how light goes and how clocks move and tick. Note that we have, in a certain sense, gone around in a circle. We originally introduced the light-pulses and clocks as instruments which were to be used in the physical world to make measurements about space and time. These instruments then led to our introduction of the interval. Now the interval, originally some complicated number computed from the measurements made with our instruments, turns out to actually be enough to determine how these instruments act. One thus comes to feel somewhat more secure in our attitude: What is to be known about the world is in the two books. From these books, and from the description of phenomena in terms of collections of events, predictions can be made about spatial and temporal relationships between phenomena. This attitude, if you like, is the relativity view.

6 The Physics and Geometry of the Interval

In chapter 5 we were concerned with the various physical interpretations that a local observer (the nonrelativist) might give to the time-readings of his clock. In that chapter, we were concerned particularly with interpretations which bear on the interval—its physical and intuitive significance, its meaning. The present chapter is in a sense an extension of that one. Once again, our interest centers on interpretive physical statements a local observer might make in connection with his various time-readings. Now, however, we wish to broaden the scope somewhat. We are now interested in physical statements which bear, not only directly on the interval and its meaning, but also, more generally, on various physical phenomena which may be happening in the immediate vicinity of that observer.

Fix an individual in space-time carrying one of our standard clocks. Further, fix once and for all an event p on his world-line. We pose the following question: Let this individual be an Aristotelian, that is, let him believe that events are properly characterized by the time of their occurrence and their position in space. Let this individual have access to our standard instruments: the clock he carries and light-pulses. We ask whether this individual will be successful in characterizing the events near p in the way he wants (by "time" and "position"), and, if so, what this characterization will look like.

Let q be some event near p. Then our individual can, as usual, send a light-pulse from himself to q, and receive the return light-pulse on his world-line. Using his clock he can, as usual, measure and record the two times, t_1 and t_2. He then can, as usual, substitute these times into the formulas $\triangle x = c/2\ (t_1 + t_2)$ and $\triangle t = ½\ (t_1 - t_2)$ to obtain an apparent spatial distance and an apparent elapsed time between p and q. What would our individual have to say if it turned out that $\triangle t = 0$, that is, that $½\ (t_1 - t_2) = 0$, that $t_1 = t_2$? He would interpret this as meaning that the apparent elapsed time between p and q is zero, that is, that p and q occurred "simultaneously." So, this individual would interpret some qs as being simultaneous with p (as having $\triangle t = 0$, that is, as having $t_1 = t_2$), and others as not being simultaneous. The q and the q' in figure 52, for example, might be regarded as simultaneous with p, whereas the q'' might not. Our individual might now wish to consider the collection of all nearby (to p) qs which, under the meaning above, he would regard as simultaneous with p. The result is just some collection of events in space-time. We may draw the surface in space-time through all those events. The resulting surface might be that shown in figure 53. Event q, for example, is entitled to lie on that surface, for, according to this individual, $t_1 = t_2$ for that particular event q. Note that event p itself lies on this surface (since, for $q = p$, we have $t_1 = 0$ and $t_2 = 0$, and hence $t_1 = t_2$. In the words of our observer, "Event p is

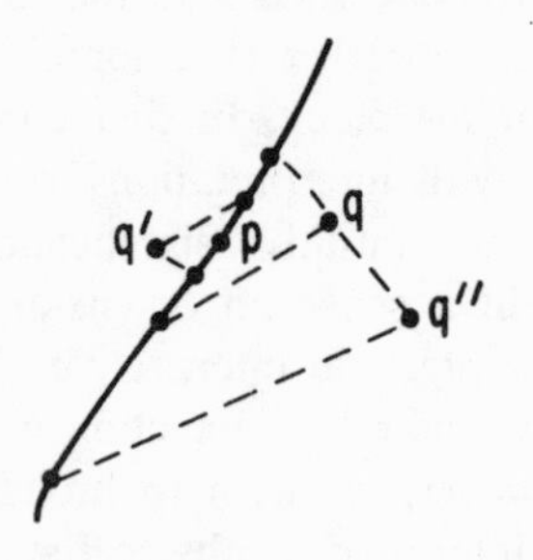

Fig. 52
Given the world-line of an observer, and an event p on that world-line, the observer can decide, for each nearby event, whether or not he regards it as simultaneous with event p.

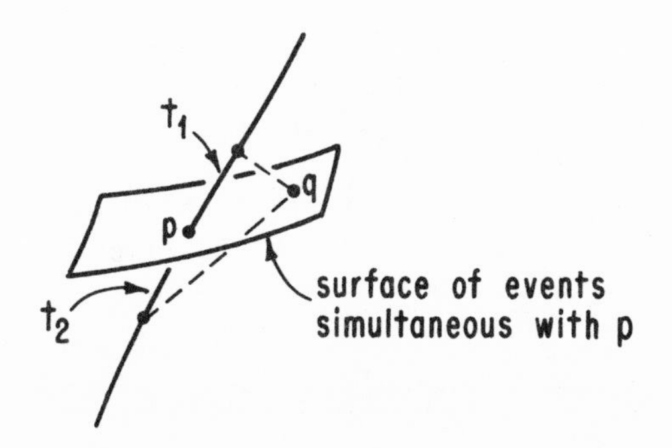

Fig. 53
The locus of events deemed simultaneous with p under the construction of figure 52 forms a 3-surface in space-time.

simultaneous with itself.") What would our individual now have to say about this surface? He would presumably say, "This surface is the locus of all events which are simultaneous with *p*. That is, this is a 'time-surface'—all its events occurred at the same time, namely, the time of event *p*."

Let us next pick, more or less at random, some time, say, 1 second. If event q should happen to be such that $\triangle t = 1$ sec, that is, such that $\frac{1}{2}(t_1 - t_2) = 1$ sec, or such that $t_1 - t_2 = 2$ sec, then our individual would of course say that q occurred 1 second after p (since the elapsed time between p and q is, for him, 1 second). We may now ask him to draw, in space-time, that surface which passes through precisely those events q with $\triangle t = 1$ sec. The result will be some other surface in space-time, say that illustrated in figure 54. Event q, for example, is entitled to lie on this surface, since for q we have $t_1 - t_2 = 2$ sec, and that is the criterion for *q*s being on our surface. This surface meets the world-line of our individual at some event, denoted by u in figure 54. What do we know about this u? Well, the surface

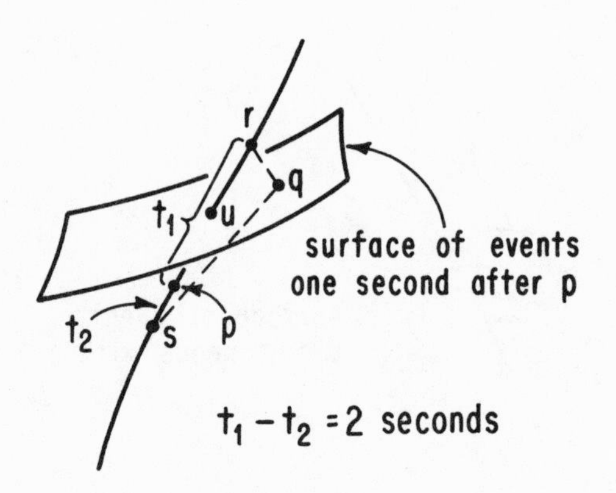

Fig. 54
The locus of events deemed to have occurred one second after event *p*.

itself is the locus of events whose apparent elapsed time from *p* is 1 second, and so, since *u* is on the surface, the apparent elapsed time from *p* to *u* must be 1 second. But both *p* and *u* are on the world-line of our observer. Thus, the elapsed time between *p* and *u* on this world-line, as measured directly by the clock on that world-line, must be 1 second. Again, we ask our individual what he has to say about this surface. He might answer, "This surface represents the locus of all events which are 1 second later than *p*, or, what is the same thing, this is the locus of all events simultaneous with *u* (since *u* is 1 second later than *p*). In other words, all the points on this surface are simultaneous with each other, and all 1 second later than the points on the first surface we drew."

We now repeat the previous paragraph, but now for other choices of the number of seconds in the first sentence: for 3 seconds, 4 seconds, 2.731 seconds, and so on. For each such choice, we will obtain a corresponding "time-surface" (fig. 55). The result is that we "slice" space-time near *p* into this family of surfaces of simultaneity. It should now be clear what is happening. These surfaces are like the old time-surfaces in our earlier discussion of Aristotelian setups. They are the "personal time-surfaces" of this particular individual. Now, however, the surfaces have been constructed with reasonable care, using our instruments.

We next do the same thing for "position in space." Consider events q for which $\triangle x = 0$, that is, for which $c/2\ (t_1 + t_2) = 0$, for which $t_1 + t_2 = 0$, for which $t_1 = -t_2$. According to our individual, these events have "zero spatial distance" from p, that is, they occurred "at the same position" as p. Again, we may draw the locus of all events q with $\triangle x = 0$ (that is, with $t_1 = -t_2$). The result will be some world-line in space-time. Our in-

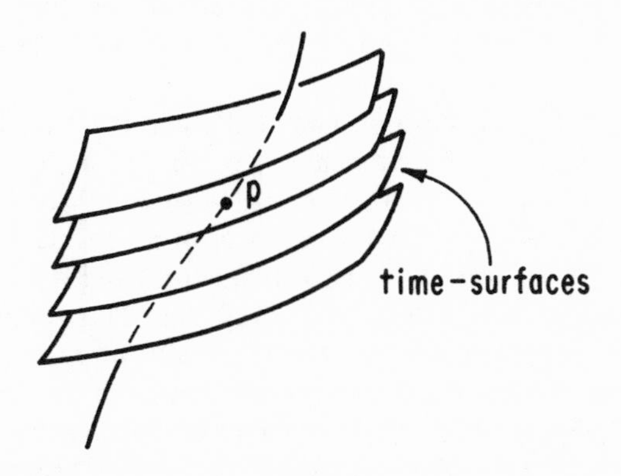

Fig. 55
Local time-surfaces as constructed by observer experiencing event p.

dividual will say of this line, "This is the locus of all events zero spatial distance from p. That is, this line represents all events at the same spatial position as p. In still other words, this is a 'position-line.' " What will this line look like in space-time? Recall (fig. 38) that t_1 is the elapsed time, according to that clock, between events p and s on the world-line, while t_2 is the elapsed time between r and p. So, since t_2 is the elapsed time between r and p, $-t_2$ is the elapsed time between p and r. But $\triangle x = 0$

means that $t_1 = -t_2$. So, in words, $\triangle x = 0$ means that the elapsed time from p to s is the same as the elapsed time from p to r. So, since both r and s are on the world-line of the clock, and since they have the same elapsed time from p according to that clock, we must have $r = s$. But $r = s$ (see fig. 49) means that event q is actually on the world-line of the clock. We summarize this discussion as follows: The *q*s with $\triangle x = 0$, that is, the *q*s with the same spatial position (according to this individual) as p, are just the *q*s which lie on the world-line of the clock. (We note in passing that this is precisely the answer we would have expected physically. If you are wandering around in the world, and some event p occurs on your world-line, say, right in front of your nose, and you are asked which events q will have the same position as p, your answer will be "those *q*s which occur right in front of my nose" [that is, those *q*s on your world-line].)

So, we have one position-line. Let's construct another. We pick some number of centimeters, say 30. Then $\triangle x = 30$ cm means $c/2\ (t_1 + t_2) = 30$, or $c\ (t_1 + t_2) = 60$, or $t_1 + t_2 = 2 \times 10^{-9}$ sec. We now ask for the locus of all events q in a certain direction (that is, the light from q is received from a certain direction), and which have $\triangle x = 30$ cm. The result will be some line in space-time. Our individual will interpret events on this line as "events which occurred 30 cm from p in such-and-such a direction." He will thus interpret the line itself as the "position-line" corresponding to the "position" 30 cm from himself in some direction. Choosing other distances and directions, we obtain a whole family of such lines in the region of space-time nearby to event p (fig. 56). In short, our individual, in this way, draws in the position-lines which are analogous to those we had in the Aristotelian setups. Again, these lines are now obtained using only the instruments we have introduced.

To summarize, an individual (characterized by his world-line) is able to make use of the instruments we make available to construct in space-time his personal time-surfaces and position-lines, which surfaces and lines cover all events near some fixed event p. The result of this construction we shall call a local Aristotelian setup ("local" because only events nearby to *p* are included). Having drawn his local Aristotelian setup, our individual is indeed able to characterize events as he wanted to in the first place. He specifies an event by giving the time-surface on which it lies,

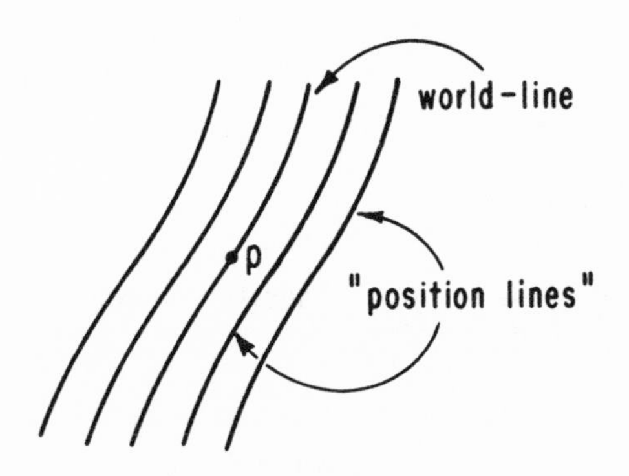

Fig. 56
Local position-lines as constructed by observer experiencing event *p*.

together with the position-line on which it lies (fig. 57). In short, he characterizes an event by its time and position.

There are two remarks to be made in connection with the discussion above. The first concerns the issue of what is actually being used in the construction. As we pointed out above, all that was used by an individual in constructing his local Aristotelian setup was two instruments—clocks and light-pulses. Recall, however, that, in the previous chapter we discovered that the behavior of these instruments could be described completely in terms of the interval. The conclusion then is that knowledge only of the interval suffices to construct the local Aristotelian setups. To put matters another way, the person in the closet with the two books will be able, given only the world-line of the individual and the event *p*, to draw in the time-surfaces and the position-lines by himself by simply consulting his books. He does not need to do any experiments; he does not need to look outside the closet.

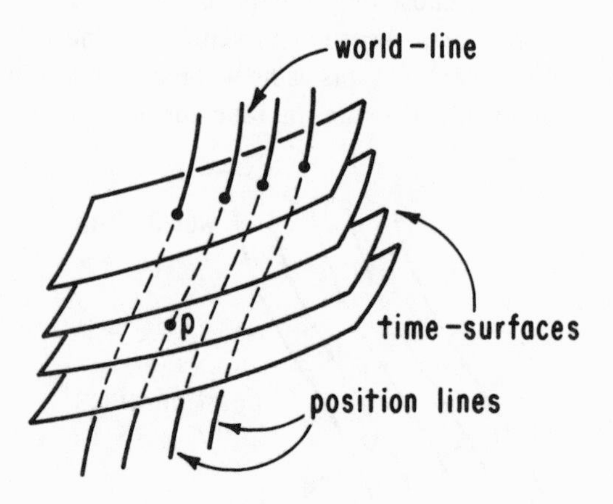

Fig. 57
The local Aristotelian setup.

The second remark is a general one regarding how, in the relativity view, individuals describe the world about them. As we have just seen, such an individual can, given an event p on his world-line, introduce a local Aristotelian setup about p. Using this setup, he can characterize events near p in terms of positions and times. Now suppose that some physical goings-on occur near p, perhaps a particle passes by, or two particles collide, or a piece of rope floats by. These things would be described, as they always are, in terms of a certain collection of events in space-time. Our individual, however, is now able to label such events by positions and times (according to him, of course). Thus, he is now able to describe these physical goings-on in everyday terms, that is, as "configurations in space which move through time." The technique for doing this is just the usual one. He intersects the world-lines, world-surfaces, and so on with his own time-surfaces. The intersection, for each such time-surface, represents for him the "spatial configuration of the phenomena at that time." Taking the intersection for successive time-surfaces, one obtains a description in terms of changes in spatial configur-ation of the phenomena with time, that is, one recovers the usual dynamical description of the physical world. Our individual has thus now been endowed with the privilege of making all the ordinary, everyday, Aristotelian statements about the world that he wishes. Of course, we are by no means offering him the guarantee that everyone else will agree with the statements he makes.

Thus, we have the following broad picture. There are events, space-time, and the intervals. Each individual constructs his local Aristotelian setup, and speaks, in terms of that setup, about the world about him. Others construct their local Aristotelian setups, and speak in their own terms about the world. The link between what one individual says and another says is this: Everybody agrees on the interval. The remainder of this chapter consists of filling out this broad picture.

In the Aristotelian view, a single Aristotelian setup is taken as "universal, intrinsic to space-time." In the Galilean view, one admits many Aristotelian setups, and, among other things, asks how these setups are related to each other. The answer in that case, remember, was as follows. All the various Aristotelians agreed on the time-surfaces, but they had different choices of position-lines. We are now in a position to ask a similar question for the relativity view. Fix an event p, and let each of two individuals experience that event (that is, physically, they pass each other at p). Each, then, may construct his local Aristotelian setup near p. We thus obtain two local Aristotelian setups in the same region of space-time, and so we may ask how they are related to each other. Let the two individuals be denoted A and B, and let their world-lines be as shown in figure 58. For the world-line A, we have already constructed the corresponding local Aristotelian setup, that illustrated in figure 57. We could now repeat this construction, but using the world-line B. That is to say, we would construct the time-surfaces and position-lines, just as before, for B. The result would be something like that shown in figure 59. Note in particular that the position-lines tend to be roughly "parallel" to the world-line, and that the time-surfaces tend to "tilt" toward the world-line. (The previous sentence is intended as just a vague, descriptive statement of roughly how the picture looks. It is in fact the way things turned out for

A, and so will presumably be how they will turn out for *B*.) We may now simply draw the composite diagram. We just superpose figure 59 on figure 57. The resulting space-time diagram is that shown in figure 60. Note what happens in the relativity picture:

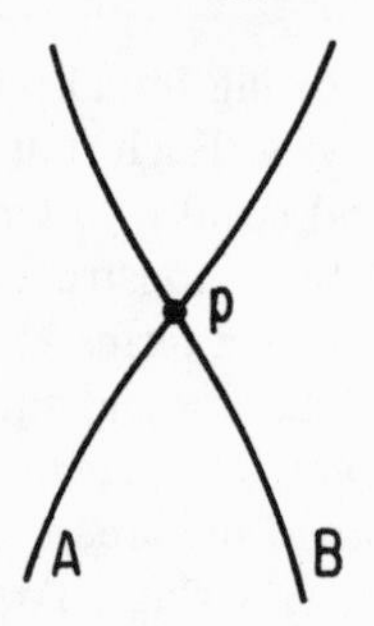

Fig. 58
Two observers, *A* and *B*, meet at event *p*.

Our two observers now not only disagree on the position-lines (something they would disagree on already in the Galilean view), but also they disagree on the time-surfaces. In short, our two observers disagree on all the usual spatial and temporal rela-

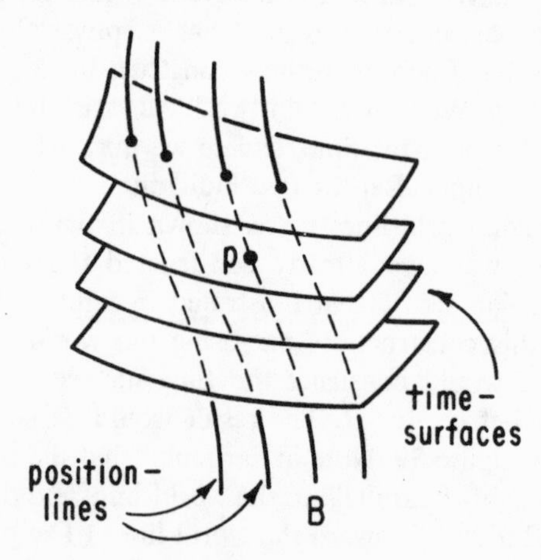

Fig. 59
The local Aristotelian setup of observer *B*.

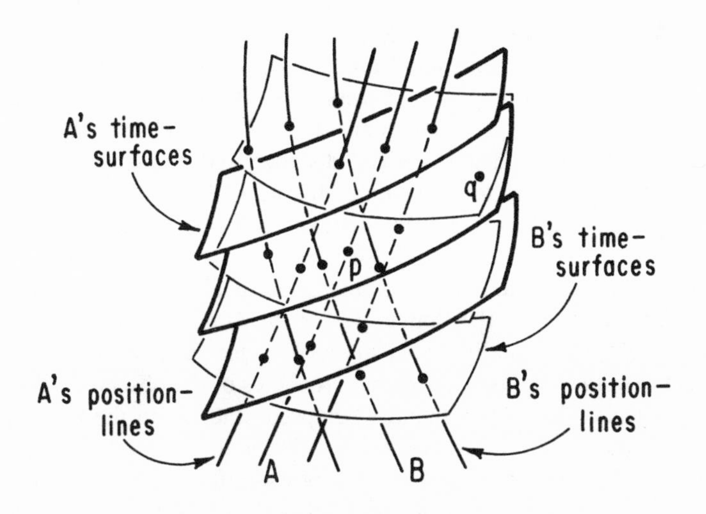

Fig. 60
Figures 57 and 59 superposed.

tions: "at the same time," "elapsed time," "at the same position," "spatial distance" for two events. What, if anything, do they agree on for two events? They agree only on the combination $[(\triangle x)^2/c^2] - (\triangle t)^2$, the interval. Consider, as an example, the event q in the diagram. Observer A would note that event q occurs on his time-surface passing through event p. Thus, observer A would say that events p and q occurred simultaneously. Event q occurs, however, not on the time-surface of B which passes through p, but rather on one of B's time-surfaces which is above B's time-surface through event p. Thus, B would claim that event q occurred after event p. Still other observers (not drawn) might claim that q occurred before event p. Thus, on the time-ordering of these two events there is no agreement. (Note that p and q in this example are spacelike related. Were they timelike related, there would indeed be agreement on which came first.) According to observer B, event q'' in the diagram occurs at the same spatial position as p (for q'' is on B's world-line, just as p is). Observer A, however, claims that q'' has a different spatial position from p.

We may now understand the transitions from Aristotelian to Galilean to relativity as a succession of decisions about what is "universal." In the Aristotelian view, all spatial and temporal relations—in particular, spatial distance and elapsed time—are taken as intrinsic to space-time. In the Galilean view, "spatial distance" between two events is taken to be an attribute of how the particular observer happens to be moving, while "elapsed time" between the events is universal to all. (It is as though "elapsed time" is the "interval" for the Galilean view.) In the relativity view, finally, both "spatial distance" and "elapsed time" between two events are personal attributes of the observers. What is universal is not the quantity "elapsed time" as in the Galilean view, but rather the combination which is the interval.

We now wish to give a number of examples of these "personalized interpretations" and the relationship, via the interval, between them.

Let us first consider the following experimental situation. One individual, A, stands around holding one of our standard watches. A second individual, B, also carrying one of the standard watches, moves past A, passing A at event p. (Of course, when we say "moves past A" we are rather speaking from A's point of view. Individual B would say that A "moves past B," passing him at event p.) This experiment is to take place in a well-lighted room. In particular, then, individual A can look at (that is, receive the light emitted from) B's watch as B approaches A, passes A, and then recedes from A. Since A can now follow continuously (through the received light) the ticking of B's watch, he can compare it with his own. We ask, How will the ticking rate of B's watch (as seen by A through the light from B's watch he receives) compare with the ticking rate of A's watch (as seen directly by A, since A is holding his watch)? The first step in answering this question (and, indeed, in answering essentially any question in relativity) is to draw a space-time diagram. (It should be noted—and this is important—that we have here specified the experiment in sufficient detail to enable us to draw the diagram. Were some de-

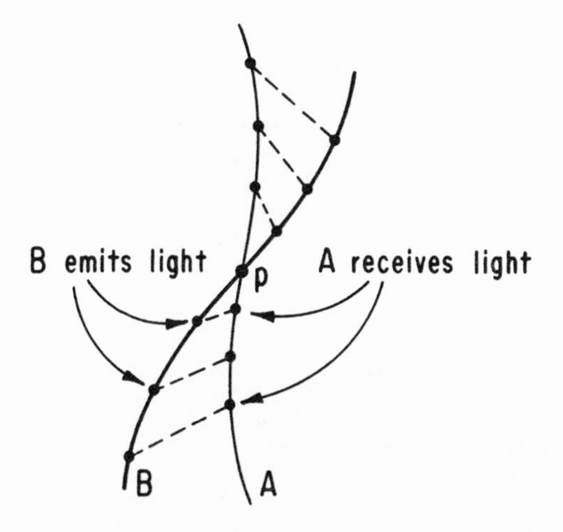

Fig. 61
Two observers, *A* and *B*, meet at event *p*. Observer *A* receives light from *B*, by means of which *B*'s clock-readings are transmitted to *A*.

tails [for example, how *A* goes about reading *B*'s watch] omitted, they would have to be supplied before one could even attack the question at all.) The appropriate space-time diagram is that shown in figure 61. *A* and *B* are each represented by a world-line, and the lines meet at event *p* ("passing"). The various events on *B*'s world-line correspond to various time-readings on *B*'s clock. This information is to be carried to *A* via light-pulses, and so these light-pulses are drawn in. Finally, the events on *A*'s world-line correspond to the receipt of the light from *B*'s watch. (It is crucial that all the relevant physics be included in the diagram, including, in this case, the light-pulses by which *A* receives his information.) Let us now redraw the diagram (see fig. 62), keeping just the features that we shall need for the calculation. We fix some event *q* on *B*'s world-line. Event *r* on *A*'s world-line is so adjusted that light just passes from *r* to *q*, and event *s* represents the return signal from *q*. To make the situation explicit (we shall do it more generally in a moment) we have inserted some

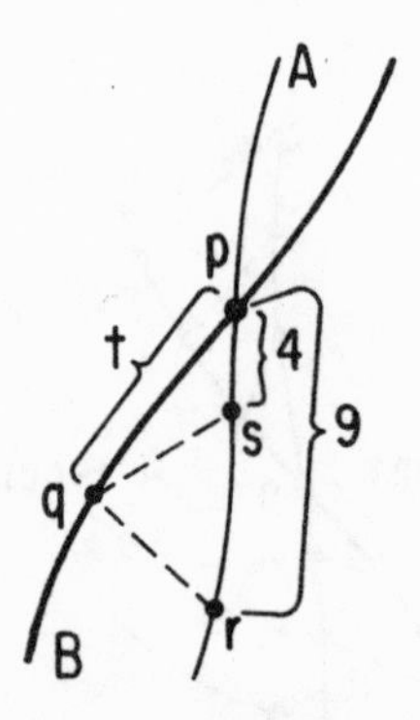

Fig. 62
The space-time diagram for the comparison of *B*'s clock as seen by *A* with *A*'s clock.

numerical values for *A*'s clock-readings. We suppose that *A* records 9 seconds between events *r* and *p*, and that *A* records 4 seconds between *s* and *p*.

It is what *A* has to say that counts in this problem, so let us see what *A* would have to say about what we have done so far. He would say, "At event *r* I emitted light which went over to person *B*. I received the return signal 5 seconds later (event *s*). Then, 4 seconds later (event *p*), person *B* passed me." Now, at event *s* person *A* received the light which was emitted from *B* at event *q*. Thus, at *s* person *A* will see *B*'s clock reading whatever it was reading at event *q*. At event *p*, on the other hand, *B* was passing *A*. Thus, at *p* person *A* will see *B*'s clock reading whatever it was reading at event *p*. Between these two events—*s* and *p*—person *A* measures an elapsed time of 4 seconds. How many ticks will *A*, between events *s* and *p*, feel that *B*'s clock made? Since at *s*, *A* sees the reading of *B*'s clock at event *q*, and at *p*, *A* sees the reading of *B*'s clock at event *p*, *A* will see, between *s* and *p*, *B*'s clock make as many ticks as *B*'s clock makes between events *q* and *p*. This, then, is the number we must compute: the elapsed time according to *B*'s clock between events *q* and *p*. This number is denoted by *t* in the diagram.

To carry out the calculation, we focus on the interval between events p and q. Let's first carry out the calculation of the interval from A's point of view. For A, we have just the usual setup (fig. 38): A sends light to q, receives the return signal, and wants to know the interval between p and q. He does the usual calculation. For t_1 (the elapsed time between p and s) he obtains -4; for t_2 (the elapsed time between r and p) he obtains 9. So, for the interval between p and q he obtains $t_1 t_2 = (-4)(9) = -36$ sec². This, then, is the interval, as determined by A. We now calculate the interval from B's point of view. But B directly experiences both events p and q. For him, therefore, the interval is just minus the square of the elapsed time he measures (directly) between q and p. That is, for B we have interval $= -t^2$.

Thus, both A and B have now computed the interval between p and q (Note: between the same two events). Person A obtains -36 sec² and B obtains $-t^2$. But the interval between two events is supposed to be universal in relativity; everybody is supposed to obtain the same answer. Hence, we must have $-t^2 = -36$, or $t = 6$. We have now answered our question. The elapsed time that B measures (directly) between events q and p is just 6 seconds.

So, what is happening physically? In the stretch between s and p, A experiences an elapsed time of just 4 seconds. On the other hand, A sees, during this stretch, B's clock tick through 6 seconds (for the elapsed time, according to B, between q and p, is 6 seconds). Thus, A sees B's clock tick through 6 seconds during only 4 seconds of "real time." In other words, A sees B's clock as speeded up by 50% (6 ticks in 4 "real" seconds).

Note how one solves problems of this kind. One first describes the situation physically, that is, one decides what one wants to ask. One then draws a space-time diagram, including all physical phenomena. One then introduces values for the known quantities (in this case, A's time-readings), and unknowns for the unknown quantities. One then searches for two events such that more than one person can compute the interval between them. (Some of the "computations" involve known quantities, some unknowns.) Finally, one sets the expressions for the interval equal to each other, and solves for what unknowns one can. Once the unknowns become knowns, one can reinterpret the situation physically. One must carefully describe the physics at the beginning and the physics at the end, and one must keep in mind exactly what is happening physically throughout the discussion.

As long as we have figure 62 before us (now, with $t = 6$ sec), let us compare a few other things.

What would person *B* have to say about the ticking rate of *A*'s clock? To answer this question, we must decide exactly what *B* is going to measure, and how he is to interpret his measurements. Let us allow *B* to do the experiment on *A* that *A* just did on *B*. At event *q*, then, person *B* saw the light emitted from *A* at event *r*. In other words, at event *q* person *B* sees *A*'s watch reading whatever it was reading at event *s*. At event *p*, on the other hand, person *B* sees *A*'s watch reading whatever it was reading at event *p*. In other words, the description according to *B* is the following: "At event *q*, I saw *A*'s watch reading something (namely, whatever its reading was at *r*). Then, 6 seconds later, person *A* passed me (event *p*). During these 6 seconds, I saw *A*'s watch undergo 9 ticks (since *A*'s watch measures 9 seconds between events *r* and *p*). So, I saw *A*'s watch make 9 ticks in 6 'real' seconds, that is, I saw *A*'s watch speeded up by 50%."

To summarize, *A* says that *B*'s watch is speeded up by 50%, while at the same time *B* says that *A*'s watch is speeded up by 50%. Who is right? Who is really speeded up and who isn't? The answer to this question is that there isn't any "really"—in relativity or in physics. One lives in the world, one does observations on various things, and one interprets the results of those observations. Nobody ever promised that there wouldn't be occasional (or even frequent) disagreements. The idea that someone must be "really" speeded up and someone must be wrong is a holdover from the Galilean (or even Aristotelian) view. The conclusion may perhaps be a bit surprising, but it is certainly not "logically inconsistent" in any sense, and, as we shall see in a moment, it is not even in disagreement with what we observe in everyday life.

Let us next, still using figure 62, compute what A would say that B's speed is. (Person A will, of course, here use his local Aristotelian setup.) So, person A will note that B directly experienced both events q and p. Person A may now compute the apparent (according to him) spatial distance and apparent elapsed time between these two events. This is easy, for he has for these two events $t_1 = -4$ sec and $t_2 = 9$ sec. So, for spatial distance, we have $\triangle x = c/2\ (t_1 + t_2) = c/2\ (-4 + 9) = 2\frac{1}{2}\ c$. For elapsed time, we have $\triangle t = \frac{1}{2}\ (t_1 - t_2) = \frac{1}{2}\ ((-4) - (+9)) = \frac{1}{2}\ (-13) = -6\frac{1}{2}$ sec. Why is the apparent elapsed time between p and q negative? Because, according to person A, event q actually occurred before event p. Thus, according to A, person B directly experienced two events having a spatial separation of $2\frac{1}{2}\ c$ cm and an elapsed time of $6\frac{1}{2}$ sec. (We may reverse the sign in the latter by considering the two events in the proper order: first q and then p.) Thus, A would say that "B moved a spatial distance of $2\frac{1}{2}\ c$ cm in $6\frac{1}{2}$ sec." To compute the speed of B, then, A would divide the distance by the time, that is, he would have the speed of B (apparent to A) $= (2\frac{1}{2}\ c)/(6\frac{1}{2}) = 5/13\ c$. That is to say, A says that B is moving at 5/13 (about 38%) of the speed of light. In centimeters per second, this speed is $5/13\ c = 5/13\ (3 \times 10^{10}) = 1.2 \times 10^{10}$. In short, A, in our numerical example, feels that B is moving by at an incredible speed—38% of the speed of light. Such large speeds, large fractions of the speed of light, are in fact necessary to obtain large disagreements (in this case, 50%) in the apparent clock rates. The fact is that in everyday life one does not see one's friends walk by at 38% of the speed of light —and so one does not see one's friends' clocks speeded up by 50%.

One might well suspect that all this has been "prearranged" by the original numerical choices of times—4 seconds and 9 seconds. To allay such suspicions, we very briefly indicate how the calculation would be done in general. Let us replace "9" by some general number, a, and "4" by b. Then quite generally a will be greater than b, since r certainly occurs before s, according to A, in the diagram. Then we shall have $t_1 = -b$ and $t_2 = a$, and so the interval between p and q, according to A, will be interval $= t_1 t_2 = -ab$. According to B, interval $= -t^2$. Equating intervals, we have $-t^2 = -ab$, and so $t = \sqrt{ab}$. This is the formula for B's elapsed time between q and p. Dividing B's elapsed time between q and p by A's elapsed time between s and p, we obtain the factor by which A regards B's clock as speeded up. The result is $\sqrt{ab}\ /\ b = \sqrt{a/b}$. Since a is greater than b, this factor will be greater than 1, and so A will always feel that B's clock is speeded up. Dividing A's elapsed time between r and p by B's elapsed time between q and p, we obtain the factor by which B regards A's clock as speeded up. The result is $a/\sqrt{ab} = \sqrt{a/b}$. Again, since a is greater than b, B always feels that A's clock is speeded up. Note that the factors in these two cases are the same, in agreement with our earlier numerical calculation. Next, let us compute the speed, according to A, at which B passes by. For the spatial distance, we have $\triangle x = c/2\ (t_1 + t_2) = c/2\ ((-b) + a) = c/2\ (a - b)$. For the elapsed time, $\triangle t = \frac{1}{2}\ (t_1 - t_2) = \frac{1}{2}\ ((-b) - (+a)) = -\frac{1}{2}\ (a + b)$. We reverse the sign of $\triangle t$, once again, to get the events in the correct order. Then, for the apparent speed according to A of B, we obtain apparent speed $= [c/2\ (a - b)]/[\frac{1}{2}\ (a + b)] = c\ (a-b)/(a+b)$. Thus, A thinks that B is going at a fraction $(a-b)/(a+b)$ of the speed of light. (We might note in passing that $(a-b)/(a+b)$ is always less than 1, and so A always thinks that B is going less than the speed of light.) We may easily see, from this calculation, what happens in the "everyday" case. If A is to see B going by at some "reasonable" speed, that is, at a very small fraction of the speed of light, then it will be necessary for $(a-b)/(a+b)$ to be a very small number. This can be accomplished only if $(a-b)$ is very small, that is, only if a is practically the same as b. But, for a and b nearly equal (say, $a = 4.000000001$ and $b = 4$), the "speeding-up factor," $\sqrt{a/b}$, will be very nearly 1 (1.0000000001 in our example). In short, for "reasonable" speeds, each A and B see the other's clock going at practically the "right" rate. This, then, is the reason why we don't, in everyday life, argue about the accuracy of each other's clocks.

So, we have now asked our physical question, and we have answered it in detail. Let us now return to *A* (reinserting the numerical values), and ask him what he thinks about all of this. He might reply, "Our conclusion is that I see *B*'s clock as speeded up, for I see it undergo 6 ticks in 4 seconds. All this is not the least bit surprising, however. I recorded my 4 seconds between *s* and *p*, whereas *B* recorded his between *q* and *p*. The fact is (and here *A* is talking a bit like a nonrelativist) that *B* started his stopwatch at *q*, while I had to wait for the light from *q* to finally reach me at *s* before I started my stopwatch. We both stopped our stopwatches at *p*. Of course *B* should record a longer time than I, because I had to wait while the light was coming to me before I could start my watch. The whole comparison isn't fair. If we really want to make a proper comparison of clock-rates, I should be allowed to start my watch 'at the same time' as *B* does." We need have no concern with judging *A*'s sense of fair play. All we need get from *A* is a clear statement of what new experiment he wishes to do. Then, we may analyze what will happen. Let *A* decide on the following experiment (fig. 63). Person *B* will, as always, measure the elapsed time between *q* and *p*, and he will, as before, obtain 6 seconds. Now, however, *A* will start his stopwatch earlier. He will now measure, not the elapsed time between *s* (the event of his seeing the light from *B*) and *p*, but rather the elapsed time from some event *u* to *p*. This event *u* will be so selected by *A* that he will regard it (according, naturally, to his local Aristotelian setup) as simultaneous with event *q*. Now, we wish to compare *B*'s elapsed time (6 seconds) with *A*'s elapsed time from *u* to *p*. (We note that this new experiment is much less "directly observational" than the earlier one. Now, *A* does not just compare the rate he sees *B*'s watch ticking with his own rate. Rather, he computes the location of event *u* [a computation based on his own sense of fair play], and compares his elapsed time from *u* to *p* with *B*'s from *q* to *p*.)

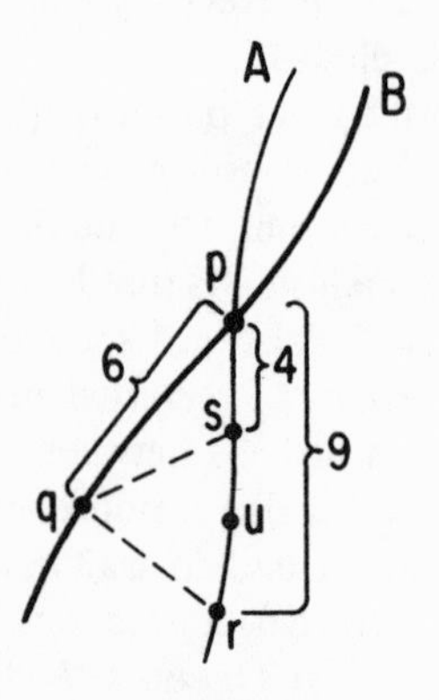

Fig. 63
A second version of the experiment of figure 62. This time, A compares B's apparent elapsed time between q and p with A's elapsed time between event u (deemed by A simultaneous with q) and p.

So, we must figure out where u is to be. According to A's local Aristotelian setup, the elapsed time from q to p is 6½ sec (for $\triangle t = \frac{1}{2}\ (t_1 - t_2) = \frac{1}{2}\ ((-9) - (+4)) = -\ 6\frac{1}{2}$ sec, a calculation we did earlier). So, if A wishes to regard u as simultaneous with q, he must choose u to be 6½ sec before p on his world-line.

Now, A is satisfied that he gets a fair break. What is the situation? Now, A records an elapsed time of 6½ sec from u to p, while B records an elapsed time of 6 sec from q to p. Thus, A would say that B only recorded 6 sec in 6½ "real" seconds. In other words, A would now say that B's clock is slowed down by about 8%. This, then, is our result for this new experiment.

What in the world is going on here? First we decide that B's clock is speeded up by 50% relative to A's, and then we decide that B's clock is slowed down by 8% relative to A's. What is the right answer: speeded up or slowed down? 50% or 8%? The answer to this question is, in my view, the crux of this whole subject. Consider for the moment the words *speeded up* and *slowed down*. They are everyday words in our everyday language; they arose from the context of the world in which we live. Our mental picture of a clock that is "speeded up" is of a clock that somehow doesn't work properly; its internal mechanism has somehow been adjusted so that it runs "too fast." In short, our picture is essentially a Galilean (more likely, an Aristotelian) one. There are many different ways in which one could actually determine experimentally whether one clock is "speeded up" or not. One could hold it next to a normal clock; one could look at it at a distance, comparing with a normal clock; one could bounce light-pulses around and make it a "fair comparison" as in the second situation above; and so on and so on. There are a vast number of different methods of comparison. Now in everyday life all of these different comparison methods go under the single term *speeded up*. The reason, of course, is that they all give the same answer in everyday life. There is no need to invent different words to describe the different comparison methods when, since the normal results of different methods are the same, a single *speeded up* will do. In relativity, on the other hand, different comparison methods will normally yield different answers. The single, inclusive, term *speeded up* will just no longer suffice. If you want to know what will result, you have to describe in considerably more detail than just "Is it speeded up or not?" what actual experiment you propose to do. Given a statement of the experiment in sufficient detail, relativity will produce a prediction. The Eskimos, so I understand, have over twenty-five different words for *snow*, words which distinguish the various subtle differences between various types of "snow." We have just one word. If light moved much more slowly in everyday terms then, I can guarantee, we would have something like twenty-five different words for *speeded up*. Similar remarks would apply to many other terms, including "speed of travel," "at rest," "simultaneous," "elapsed time," "spatial distance," "same position," "length," "straight," and so on. For these and many other everyday terms, one must take extreme care, in relativity, not to use them thoughtlessly.

Before leaving the present example, we make one further observation. We have, for all practical purposes, actually seen the space-time diagram of figure 63 before; it is essentially the same as figure 36. The world-line A in figure 63 corresponds to the A in figure 36. The stretch of B between q and p in figure 63 corresponds to the stretch of B between s and q in figure 36. Event u in figure 63 corresponds to event r in figure 36. Our conclusion in figure 63 is that A's elapsed time between u and p is greater than B's between q and p. Translating to figure 36, we would have that A's elapsed time there between r and q is greater than B's between s and q. But this is precisely the experimental result of the experiment depicted in figure 36. The mu-meson internally feels that it experiences a rather small elapsed time, so that it does not feel obligated to decay. Thus, we are now able to explain the mu-meson experiment in terms of relativity. In a sentence, the explanation is that "Everyone agrees on the interval only, with the result that the prediction is essentially what is observed experimentally." This observation, then, is intended as one piece of evidence that we are going in the right direction.

One might well object at this point that we have not "explained" anything. We have merely replaced one baffling experimental observation, concerning mu-mesons, by another equally baffling assertion, concerning the interval. Why do we not "explain" why the interval should be the same for everybody? What is the real reason? The best answer I can give is that physics never finds any final explanations or real reasons. Rather, some things are merely related to other things, and those to other things, and so on. One aims for a conceptual simplicity and some feeling of an understanding of how the world works. One does not, as a physicist, know how to do better; one becomes accustomed to accepting gratefully those small glimpses into the workings of nature that one is able to glean.

We now turn to a second example. Now, observer A decides to emit a light-pulse, and decides to measure its speed as the light recedes from him. If this is going to be all observer A chooses to tell us about what he is to do, we are helpless. We must ask him for more details about his actual experiment. He provides the following details. At event p, he will emit a light-pulse. A second event, q, along the world-line of the pulse will be chosen. Person A will then proceed to construct his local Aristotelian setup about p, and, in particular, will be able to determine his personal elapsed time and spatial distance between events p and q. He will then divide the apparent spatial distance by the apparent elapsed time to obtain the apparent speed. We may now compute what his answer will be (fig. 64). Let a return

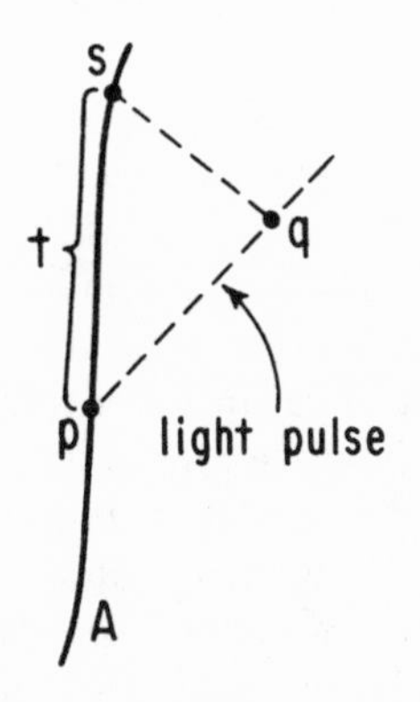

Fig. 64
Space-time diagram of an experiment in which A determines the apparent speed of light.

light signal from event q reach his world-line at event s. Let the elapsed time, according to A, between events p and s on his world-line be t. We wish to compute the apparent spatial distance between p and q, as well as the apparent elapsed time, both in terms of this number t. We recognize this situation as the usual one (fig. 38). In this case, event r is the same as event p. The corresponding times are $t_1 = t$ and $t_2 = 0$ (the latter, since $r = p$). We therefore just compute spatial distances and elapsed times in the usual way. For the spatial distance between p and q, we have $\triangle x = c/2\ (t_1 + t_2) = c/2(\ t + 0) = \frac{1}{2}\ ct$. For the apparent elapsed time between p and q, we have $\triangle t = \frac{1}{2}\ (t_1 - t_2) = \frac{1}{2}(t - 0) = \frac{1}{2}\ t$. Thus, person A would say that the light appeared to travel spatial distance $\frac{1}{2}\ ct$ in time $\frac{1}{2}\ t$. So, he would obtain speed $= (\frac{1}{2}\ ct)/(\frac{1}{2}\ t) = c\ (= 3 \times$

10^{10} cm/sec). This, then, is the answer: He always obtains this number c.

Should we rejoice at this result, claiming now to have also explained the experimental observations of constancy of the speed of light discussed in chapter 10? Most certainly not. We have, it is true, asked a definite experimental question, and obtained an answer. We must now decide whether our experimental arrangement is likely to be the one alluded to in chapter 10. To this end, let us describe the present arrangement in a little more detail. The only thing that A actually measures is the single elapsed time t; everything else is computation. (One might already be a little suspicious, for the t drops out of the computation.) What, then, does A do with this single number t? First, A sent light out at event p, and received it back at s. Since s was, for A, t seconds after p, person A regards q as having occurred ½ t seconds after p. This is how we arrive at $\triangle t$. Further, A notices that the total round-trip light-travel time from himself to q is t seconds. He regards the light, therefore, as having traveled a total spatial distance of ct (where, of course, $c = 3 \times 10^{10}$, a number, say, that he looked up in the encyclopedia). Since the light made a round trip, he regards q as spatial distance ½ ct away. This is how we arrive at $\triangle x$. Now it should be obvious that when you divide one by the other you are going to get c. The fact is that "the speed of light is c" was already put into the calculation at the beginning, and just came back out again at the end. We have not really computed anything. To put it another way, if A published an article, describing this "experiment" and claiming thus a demonstration that the speed of light is the very same c that he found in the encyclopedia, people would laugh at him. Now this fiasco is not the fault of relativity. It is all A's fault. Person A told us what the experiment was to be, and we told him the (correct) answer. Relativity ought to give the correct answer even for silly experiments.

Let us see, therefore, if we can design a more reasonable experiment for measuring the speed of light (recalling that we shall have to do it in some detail, since *speed* may have many different meanings, depending on the experiment; one "meaning," for example, is that above, a rather uninteresting meaning in the present case). It should be clear, first of all, that we shall never find a better experiment if we restrict ourselves to just clocks and light-pulses. The only numbers we can obtain from these instruments (without actually using *c* itself) are numbers of seconds. But whatever speed we finally obtain ought to be a number of centimeters per second. To obtain such, we shall need something that records centimeters, say, a meter stick. Doesn't this violate our original precept about using just two instruments, clocks and light-pulses? Some explanation is required. The decision to use just the two instruments was made when we were trying to decide what was to be intrinsic to space-time. Not wanting to be confronted with a great variety of instruments, we selected just two. We have now, however, introduced and discussed the interval: We are done with that part. Our selection of instruments to use in measuring the interval was not intended to suggest that the world, according to relativity, would never contain anything except clocks and light-pulses. There are many other things in the world. Such "other things" represent perfectly respectable physical objects. The making of measurements with these other things is a perfectly respectable class of physical phenomena. As such, these phenomena, too, should be representable within space-time. In short, it should be possible in principle to insert into space-time any measuring instrument. Our choice of clocks and light-pulses as the basic instruments was based primarily on the simplicity of their representations.

One typical experiment, then, might be the following. We build standard meter sticks. Two individuals meet at the center of the meter stick, and there synchronize their watches. These individuals then separate, one going to each end of the meter stick. Having reached their respective positions, the individual at one end of the stick emits a pulse of light down the meter stick, recording the time of emission on his watch. The individual at the other end records the time of receipt of the light-pulse. Finally, the two individuals meet again at the center of the meter stick, and compare their watches. One then takes the total time between the emission of the light and the final meeting at the center according to the emitting individual minus the total time between the receipt of the light and the final meeting at the center according to the receiving individual: this difference one regards as the "total time of flight" of the light. Finally, one divides 100 cm (the length of the meter stick) by this total time of flight to obtain a speed for light. This, then, is a typical experiment. Its space-time diagram is shown in figure 65. The world-lines of the

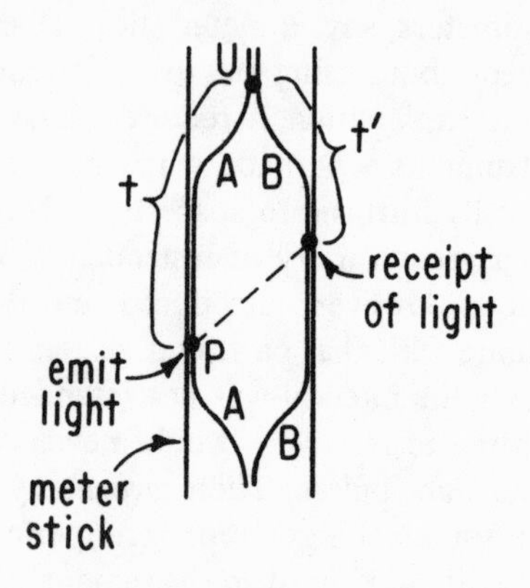

Fig. 65
Space-time diagram for an experiment in which the speed of light is measured directly, using a meter stick and clocks. Two observers, *A* and *B*, synchronize their watches, separate to opposite ends of the meter stick, and time the passage of light across the stick.

two individuals (*A*, the emitter, and *B*, the receiver) are shown. The formula for the apparent speed of light is $100/(t - t')$.

One might imagine that, at this point, the experiment has been described in sufficient detail that one ought to be able to carry out some calculation to determine whether or not the speed that will be obtained will in fact be what we expect, 3×10^{10} cm/sec. In fact, however, a number of important details have been omitted. For example, we have not specified in detail exactly how individual *A* is to travel from event *p* to event *u*. Indeed, we know, from our earlier discussion, that the value of *t* we obtain (and hence the value for the speed of light) will depend on his path in space-time. Similarly, we have not specified how *B* travels. Even more seriously, we have not described the meter stick in sufficient detail. If, for example, the meter stick were made out of rubber, so that it could be stretched or bent, then we would expect to obtain the wrong answer for the speed of light. But how can one tell whether or not the meter stick has been stretched; how can one tell that the distance between successive marks on the stick remains 1 centimeter? Comparison with a second meter stick is no good, because we might have the same doubts about the second stick. Measurement of an apparent spatial distance between successive marks (using light-pulses) is also unsatisfactory, for implicit in such a measurement would be the value 3×10^{10} cm/sec for the speed of light. In short, we must apparently introduce some sort of check that we have a true meter stick in our experiment, and such checks seem to be difficult to describe.

It is, as it turns out, possible to deal with all of these various difficulties. The result is a highly technical experiment. Its physical description is a long story, and its incorporation into space-time is a rather complicated business. For this reason, we shall not attempt to carry out this program in any further detail, but rather will merely quote the result that would be obtained. That result is, of course, that everybody obtains for the speed of light 3×10^{10} cm/sec.

One might very well be unhappy at this turn of events. The "constancy of the speed of light" was one of the features that helped upset the Galilean view. How can we allow it to be such a complicated business in the relativity view? How can all this be squared with the fact that many elementary textbooks in relativity contain sentences of the form "The second postulate of relativity is that everybody obtains for the speed of light 3×10^{10} cm/sec"? My response would be that (1) It is indeed unfortunate, and perhaps even a reasonable cause for unhappiness, that certain observations which historically led to the formulation of a theory turn out to be complicated within the final theory. However Nature, understandably I suppose, seems not to care at all how simple or complicated these historical observations are in the final theory. The fact is that it seems necessary to use meter sticks to measure the speed of light, and meter sticks, whether we like it or not, are rather complicated in relativity. (2) In my opinion, any statement of the form "The nth postulate of . . . (some physical theory) is . . . (some brief statement)" is, at the very least, suspect, and, at the very most, a waste of time. I see physics as an ongoing process of acquiring an ever deeper understanding of nature—everything is tentative, confusing, subject to caveats and interpretation. I just do not believe that "postulates" are relevant to physical theories. (3) Even if one insists on postulates, why not choose for one's "postulates" things which fit simply and naturally into the final theory (such as statements about the interval) rather than complicated, historically based observations?

We now turn to another class of observations, namely measurements of the length of an object such as a long pole. To fix ideas, let us set up the following situation. An individual, A, is standing around with his standard watch. Our pole comes at him "head-on." First the front end of the pole passes him, then the body of the pole, and finally the rear end of the pole. This is shown in figure 66. Note the world-surface of the pole, the world-lines of the front and rear ends, and the events at which A meets those two ends. Our individual A decides to determine for him-

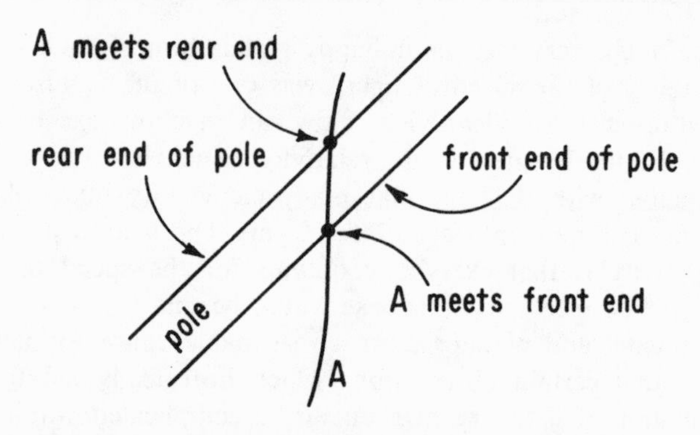

Fig. 66
Space-time diagram of observer A being passed by long pole.

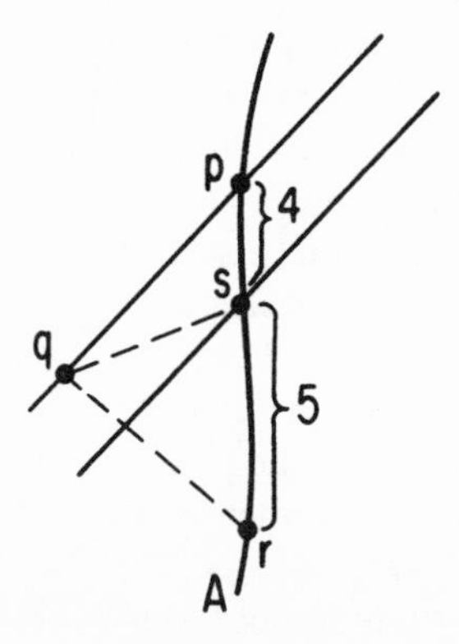

Fig. 67
One method for *A* to determine the "length" of the pole. *A* computes the apparent speed of the pole and the apparent time taken by the pole to pass him.

self the (apparent, according to him) length of the pole. As we have already remarked, we cannot deal with this situation on the level at which it has been stated so far. We require a more detailed statement from *A* of precisely what he is going to do experimentally to make his measurements. There are numerous possible experiments that *A* might do. We will consider just a few examples.

One possible experiment might be this. First, *A* computes the apparent speed of the pole (say, the speed of the rear end of the pole), in the way we have discussed earlier. Next, *A* times how long it takes the pole to pass him. Finally, *A* multiplies this apparent speed by the time, to obtain an apparent length of the pole. A space-time diagram for this particular experiment might be that shown in figure 67. We introduce event *q* on the rear end of the pole, so adjusted that a light-pulse emitted from *q* just reaches *A* as the front end of the pole does. We further introduce event *r*, so that a light-pulse from *r* just reaches event *q*. To make the example numerical, we further introduce some elapsed times, namely 4 seconds according to *A* between *s* and *p*, and 9 seconds between *r* and *p* (so the elapsed time, according to *A*, from r to s is 5 seconds). Let us now, with this experiment and with these numbers, compute the length of the pole according to A. So, A must first compute the speed of the pole. This, however, is a problem we have already solved. (See fig. 62. The world-line B in that figure corresponds precisely to the world-line of the rear end of the pole in the present case.) Thus, the speed of the pole in the present case is just the answer that was obtained earlier, namely $5/13\ c$. Next, A notices that it took the pole 4 seconds to pass him (elapsed time between s and p). So, for the apparent length of the pole, A obtains length $= (5/13\ c) \times 4 = 20/13\ c$. (In centimeters, this is $20/13 \times 3 \times 10^{10} = 4.6 \times 10^{10}$ cm—a very long pole indeed. In fact, it is so long that the implicit assumption of "nearby events" may not be valid. As usual, one need not worry about this. It is just a consequence of the fact that we wish to deal with "reasonable" numbers such as 4 seconds and 9 seconds, rather than unreasonable ones such as 4 seconds and 4.000000001 seconds.) So, in any case, we have computed the length of the pole according to this experimental prescription.

Individual A might, however, have chosen a quite different technique for measuring the length (fig. 68). He might, for ex-

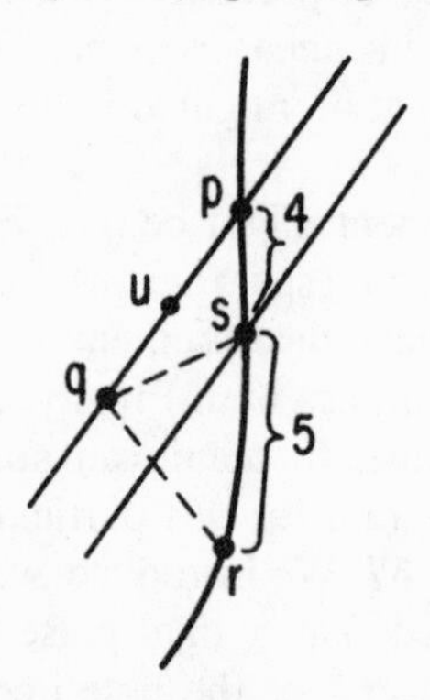

Fig. 68
A second notion of "length" of the moving pole. A determines an event u which he regards as simultaneous with event s, and then computes the apparent spatial distance between events u and s.

ample, have proceeded as follows. He might identify some event u on the world-line of the rear end of the pole which he regards (according, of course, to A's local Aristotelian setup) as simultaneous with event s. Having obtained two events, u and s, which are on opposite ends of the pole, and which A regards as simultaneous, A would then proceed to compute the apparent spatial distance between those two events. (Note that A would very much prefer that the events be simultaneous, so he can feel that "the pole hasn't moved any between u and s, distorting the measurement.") Finally, A might regard the resulting apparent spatial distance as the length of the pole. Let us now compute the length of the pole by this method. First note that, according to A, the elapsed time from s to p is 4 seconds. Hence, if u is to be simultaneous with s, then it had better be true, according to A, that the elapsed time from u to p be 4 seconds too. Thus, u is that event on the world-line of the rear end of the pole which, according to A, has an elapsed time of 4 seconds from p. But A feels that the rear end of the pole is moving at speed $5/13\ c$. So, A feels that the apparent spatial distance between u and s is just $4 \times (5/13\ c) = 20/13\ c$. (Here, we are using what we mean by "apparent speed," namely apparent speed = apparent spatial distance/apparent elapsed time.) So, we now have our two events, u and s, which A regards as simultaneous, and we have A's apparent spatial distance between them, namely $20/13$ c. So, by this method, A would regard the pole as having length $20/13\ c = 4.6 \times 10^{10}$ cm. We note in particular that this second method gives the same answer as the first (and, indeed, it probably became obvious, somewhere during the calculation above, that this was going to happen).

Let us now consider, finally, a third method for measuring the length (fig. 69). This time, individual A identifies the event, v, at which the light he sent to the rear end of the pole meets the front end of the pole. He then argues as follows. Light, beginning at event v at the front end of the pole, traveled across the pole to event q at the rear end. This light then bounced back, traversing the pole again, and returned to the front end at event s. I shall determine the apparent elapsed time (according to my local

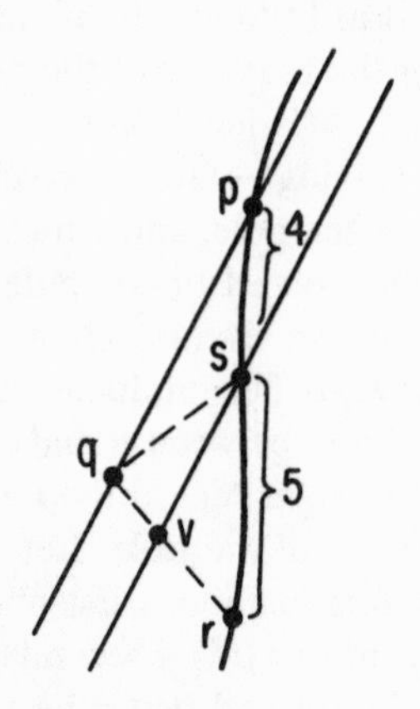

Fig. 69
A third notion of "length" of the moving pole. *A* effectively times the passage from one end of the pole (event *v*) to the other end (event *q*) and back (to event *s*).

Aristotelian setup) between event *v* and event *s*. During this time, light traveled across the pole and back. Hence, if I multiply this apparent time by *c* I shall obtain the total distance the light traveled. Dividing by 2 (since the light traveled across the pole and back), I will obtain a number which I might call the length of the pole. Let us now compute the length that *A* will obtain according to this method. Denote by $\triangle t$ the apparent elapsed time, according to *A*, between *v* and *s*. It is this number we wish to determine. We can compute it as follows. First note that, since *A* feels that the front end of the pole is traveling at $5/13\ c$, and since both event *v* and event *s* are on the front end of the pole, *A* can compute directly, in terms of the unknown $\triangle t$, the apparent spatial distance between *v* and *s*. He obtains $\triangle x$ (apparent spatial distance, *v* to *s*) = (apparent speed front end pole) × (apparent elapsed time, *v* to *s*) $= 5/13\ c \times \triangle t$. He may, however, compute the apparent spatial distance between *v* and *s* in another way. Since, according to *A*, events *r* and *s* occur at the same spatial position (namely, in the immediate presence of *A*), the apparent spatial distance between *v* and *s* (what we called

$\triangle x$ above) is the same as that between v and r. Furthermore, since the apparent elapsed time from v to s is $\triangle t$ (the unknown), and since the apparent elapsed time from r to s is 5 seconds, the apparent elapsed time from r to v must be $(5 - \triangle t)$ seconds. So, $\triangle x$ is the apparent spatial distance between r and v, while $(5 - \triangle t)$ is the apparent elapsed time between r and v. But events r and v are lightlike related; they are joined by a light-pulse. Hence, the apparent spatial distance and the apparent elapsed time between these two events must be related by multiplication by c. So, we must have $\triangle x = (5 - \triangle t)c$.

So far, we have obtained two relations, namely $\triangle x = 5/13\ c\triangle t$ and $\triangle x = (5 - \triangle t)c$. Equating the $\triangle x$s, we obtain $5/13\ c\triangle t = (5 - \triangle t)c$. Adding "$\triangle t\ c$" to each side, we have $18/13\ c\triangle t = 5c$. Dividing out the cs, we have $18/13\ \triangle t = 5$. Multiplying both sides by 13/18, we have $\triangle t = 65/18$ sec. This, then, is the apparent elapsed time, according to individual A, between event v and event s. (There must be hundreds of different ways one could do this calculation, all yielding this same result. We have here just picked one at random.)

So, now A has obtained the apparent elapsed time between v and s. All that remains is to compute the apparent length of the pole. Following the instructions he gave earlier, he first multiplies this elapsed time by c to obtain $65/18\ c$, a distance he regards as the total light-travel distance. Then, since the light went down and back up the pole, he divides by 2 to obtain $65/36\ c$, the apparent length of the pole by this method. In terms of centimeters, this is $65/36 \times 3 \times 10^{10} = 5.4 \times 10^{10}$. Thus, by this method he obtains a somewhat longer length for the pole than by the earlier methods (namely, about 17% longer).

It would not be difficult to think of ten or fifteen other methods by which A could measure the "length" of this pole. Some might give the same answer as others, but one would in general expect to obtain a variety of different answers. Probably, most would be within the general area of 5×10^{10} cm (say, between 4×10^{10} and 6×10^{10}, although there could well be a few which are out of this range). These discrepancies do not, of course, represent "errors" in any sense. Rather, they represent a concrete expression of the fact that the everyday word *length* has many different meanings in relativity. These different meanings are expressed by the details of the actual experiment to be performed. This particular pole is moving rather quickly—at 38% of the speed of light. Were the pole moving more slowly, the values would all be much closer together. Since everyday poles move rather slowly, all these methods would yield nearly the same answer, and it is for this reason that we have but a single word *length*. We emphasize again that there is nothing inconsistent or even paradoxical in all this. The fault lies not in relativity or in nature, but in ourselves (since we only look at slow-moving poles) and in our language (which, necessarily, tends to reflect the things that we see).

Let us now consider a special case of the discussion above. We shall now let the pole move, according to A, at a speed, not 38% of that of light, but rather at speed zero. We let A remain always at the front end of the pole. We are now interested in the question of what our three methods will give in this case. The first method (timing the passage of the pole) is now not even applicable, for the pole never does pass individual A in this situation. We consider the second method (fig. 70). Person A, then, locates some event s on his world-line, and finds event u on the far end of the pole, which he regards as simultaneous with s. We of course know what "regards as simultaneous" means (in A's local Aristotelian setup); namely, that the two elapsed times, both marked t in the diagram, be equal. (This is immediate from the formula for the apparent elapsed time, $\triangle t = \frac{1}{2}\,(t_1 - t_2)$.) So, A has now located two apparently simultaneous events on opposite ends of the pole. By the second method, he computes an apparent length of the pole as the apparent spatial distance between the two events u and s. Thus, length = (apparent spatial distance, u to $s = c/2\ (t_1 + t_2) = c/2\ (t + t) = ct$. (Of course, we cannot assign a numerical value unless we assign one to t.) Finally, let us consider the third method. In this method, A begins by determining the "apparent light-travel time, across the pole and back." But we can just use the same figure 70 above, for in that figure light indeed travels across the pole and back. So, the apparent light-travel time, according to A, is just $2t$.

Person A then argues that, during this time, light will have traveled distance $2t \times c = 2ct$. Finally, he argues that, since the light traveled across the pole and back, the length of the pole must be $\frac{1}{2}$ $(2ct) = ct$. We note that this is the same answer as is obtained by the second method.

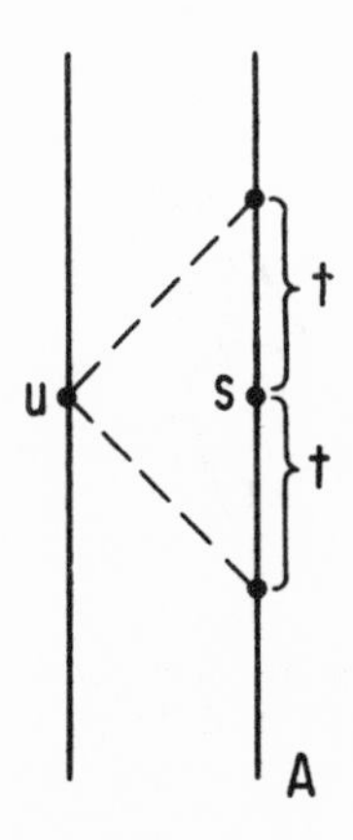

Fig. 70
A determination by *A* of the length of a pole at rest with respect to himself.

So, in the case when A regards the pole as at rest, the first method for measuring length is not applicable, and the second and third yield the same answer. (Indeed, we might have expected in this case that all applicable methods would yield the same answer, for a pole at rest is certainly "slow-moving," while in everyday life all reasonable methods do seem to yield the same answer.)

We next wish to consider a slight variation of our original problem. We introduce a second individual, B, who always remains on the front end of the pole, riding it as it passes A (fig. 71). Now, we have a single pole, but we have two individuals who are capable of observing it. Will these two individuals agree as to its length? As usual, we cannot deal with this question until we are provided with more details. In this case, we require the

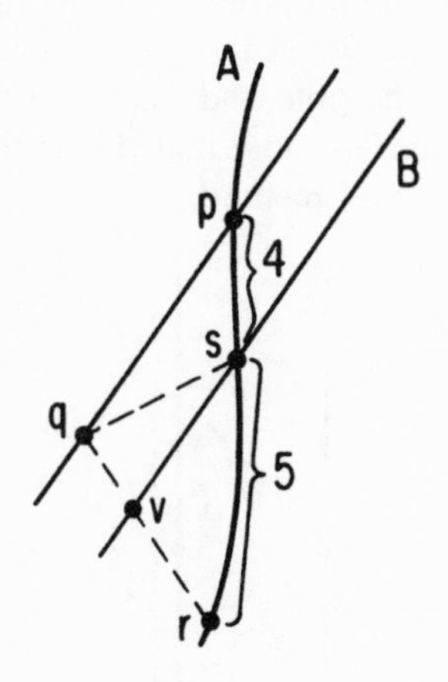

Fig. 71
A space-time diagram of an experiment in which both A and B compute the length of the pole.

details as to how A is to determine the length according to him and how B is to determine the length. (We might anticipate already that they will not normally agree, since A doesn't even agree with himself if he uses two different methods.) Clearly, there are an enormous number of possibilities here. (Let, say, there be twelve methods of measuring "length." Then there are 144 questions, each obtained by choosing a method for A and a method for B.) We don't have the patience to deal with 144 questions, so let us just pick one as an example. Let A measure the length of the pole by either method 1 or method 2 above (there being no real difference between them, since they yield the same answer; sociologically, people living in a world in which ordinary objects travel near the speed of light might have a single word, perhaps *gykwf*, for length as measured by either method). Let B measure the length by either method 2 or method 3 (which, again, give the same answer for B; of course B could not use the first method, since it doesn't work for him). So, we have now spelled out our experiment in sufficient detail to decide what the answer, according to relativity, should be; to decide whether or not A and B will agree.

The rest is fairly easy. We have already determined the apparent length of the pole according to A (using either the first or second method), namely $20/13\ c = 4.6 \times 10^{10}$ cm. We therefore have only to compute the apparent length according to B. Consider the two events v and s in the diagram. Asked about these, B would say: "At event v, I sent light down the pole. (Actually, B is taking a little undue credit here, for the light was actually emitted by A at event r, and just passed B at event v, but no matter.) This light bounced off the far end of the pole (event q), and then returned to me at event s. So, if I am to determine the apparent length of the pole according to method 3, I must find the total elapsed time that I measure between events v and s (a time that I regard as the total light-travel time across the pole and back), multiply by c, and divide by 2." So, what we must now determine is the total elapsed time, according to B, between events v and s on his world-line. This we may do, for example, as follows. We have already determined the total elapsed time, according to A, between events v and s, namely $65/18$ sec. (See discussion of method 3, fig. 69.) This, of course, will not necessarily be the elapsed time according to B. (Note that we must use B's apparent elapsed time, for B couldn't care less what A is doing. He is just using the prescribed method by himself.) However, we also obtained an expression for the apparent spatial distance between v and s according to A, namely $\triangle x = 5/13\ c\triangle t$. Since now $\triangle t = 65/18$ sec, we can substitute to obtain $\triangle x = 5/13\ c\ (65/18) = 25/18\ c$. So, we now have the apparent spatial distance and the apparent elapsed time, according to A, between v and s. Hence, the interval between events v and s, according to A, is given by $[(\triangle x)^2/c^2] - (\triangle t)^2 = [(25/18\ c)^2/c^2] - (65/18)^2 = (25/18)^2 - (65/18)^2 = -\ 100/9\ \text{sec}^2$. (It is unfortunate that these terrible fractions tend to cloud the discussion. Apparently, in such calculations, one can only avoid square roots by allowing them to appear.) So, this is the interval, according to A, between events v and s—$-100/9\ \text{sec}^2$. But the thing we are interested in is not this interval, but rather the apparent elapsed time according to *B* between *v* and *s*. Let us denote this time by *t*. Then *B* can also compute the interval between *v* and *s* using this *t*, namely, $-t^2$. So, both *A* and *B* have computed this interval. But, according to relativity, everybody is supposed to obtain the same interval. So, their values must agree, that is, we must have $-t^2 = -100/9$, or $t^2 = 100/9$, or $t = 10/3$ sec. This, then, is the elapsed time, according to *B*, between events *v* and *s*.

We now let *B* go on and compute the apparent length of the pole according to him. He multiplies this apparent elapsed time (during which, according to him, light went down the pole and back) by *c*, and then divides by 2, to obtain length $= 10/3 \times c \times ½ = 5/3\ c$. In centimeters, the length according to *B* is thus $5/3 \times 3 \times 10^{10} = 5 \times 10^{10}$.

We now have the answer to our question: *A* regards the pole as having length 4.6×10^{10} cm, while *B* regards its length as 5×10^{10} cm. They do not agree, and in fact *A* thinks that *B*'s length is too large by 8%. One might recall having seen this "8%" before. According to at least one method of comparison (fig. 63), *A* also thinks that *B*'s clocks are running too slowly by 8%. So, *A* might argue, since *B*'s times are "off" by 8% and also his "lengths" are off by 8%, *B* will probably get the same answer as *A* for the speed of light (since that speed is a ratio between apparent spatial distance and apparent elapsed time). Does this observation not therefore represent a demonstration that everybody obtains the same speed of light? I claim that in fact it does not, and that the situation here is essentially the same as that of the "silly experiment" illustrated in figure 64. Here, however, it is much more difficult to trace the argument through to see that this is the case. I leave it as a (rather difficult) exercise.

We conclude this chapter with one more example of a calculation in which different observers' observations of the world are compared. This time, we have three individuals, *A*, *B*, and *C*, so arranged that all three pass each other at event *p*. Now each of our three individuals can compute the apparent speed, according to himself, of the other two. Let us suppose, to fix ideas, that

A feels that *B* is going by at 38% of the speed of light, while *B* feels that *C* is going by at 38% of the speed of light. What we wish to compute is how fast *A* regards *C* as going by. (This experiment is already described in sufficient detail for us to calculate, for we shall agree that apparent speed = (apparent spatial distance)/(apparent elapsed time), and we know how to measure experimentally these "apparent" things.)

As a first step, we introduce light-pulses and events as shown in figure 72. Further, we let the elapsed time, according to *A*,

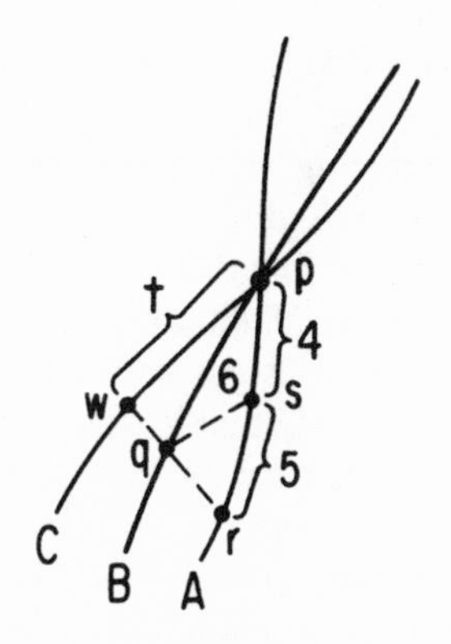

Fig. 72
Three observers, *A*, *B*, and *C*, all meet at event *p*. *A* computes *B*'s apparent speed, and *B* computes *C*'s apparent speed. What will be the speed of *C* according to *A*?

from *r* to *s* be 5 seconds, and the elapsed time, according to *A*, from *s* to *p* be 4 seconds, as usual. These choices, we know, ensure that *A* will regard *B* as going by at 38% of the speed of light. Furthermore, we have already computed (fig. 63) that, in this setup, the elapsed time according to *B* between events *q* and *p* will be 6 seconds.

The first thing we must compute is the number denoted by *t* in the figure, the elapsed time according to individual *C* between events *w* and *p*. To compute *t*, we use the fact that, according to *B*, person *C* is moving by at 38% of the speed of light. The simplest way to determine *t* is as follows. Person *A* would say,

"Person B is moving by at 38% of the speed of light. At event r I sent out light which reached B at q. Six seconds after B's receipt of the light (according to him), or 9 seconds after I sent the light (according to me), person B passed me." Person B would say (of C), "Person C is moving by at 38% of the speed of light. At event q I sent out light which reached C at event w. Just t seconds after C's receipt of the light (according to him), or 6 seconds after I sent the light (according to me), person C passed me." Since the speeds are the same in these two quotes, the ratios of the numbers should be the same. Thus, we should have $6/9 = t/6$. (The left side uses numbers from the first quote, the right side from the second.) (One could, in fact, avoid this use of ratios by introducing an extra event or two and a few extra light-pulses, letting B directly determine the spatial distances and elapsed times to find the speed of C, setting that speed equal to 38% of the speed of light, and solving still another algebraic equation. It would be a significant extra complication, and it should be more or less clear that it would yield the same result that we shall obtain by this easy method.) Solving for t, we have $t = 6 \times 6/9 = 4$ sec. So, person C experiences an elapsed time of 4 seconds between events w and p on his world-line.

What is it we are trying to compute? The speed of C according to A. We shall do this in the usual way as follows. Person C experiences directly both events w and p. So, if A wants to determine the apparent speed of C, he has only to find the apparent spatial distance (according to A, of course) between events w and p, and also the apparent elapsed time between these two events. Dividing the former by the latter, we shall obtain the apparent speed. Let us, then, denote by $\triangle x$ this apparent spatial distance (according to A, between w and p), and by $\triangle t$ this apparent elapsed time. What we wish to do is find two algebraic equations for these two unknowns.

One equation is obtained as follows. Person A can compute the interval between w and p, in terms of $\triangle x$ and $\triangle t$, in the usual way, interval $= (\triangle x)^2/c^2 - (\triangle t)^2$. Person C, on the other hand, directly experiences events w and p, and experiences an elapsed time of 4 seconds between those events. Hence, for C, the interval between w and p is $-(4)^2 = -16$ sec². But everybody is to agree on the interval between two events (in this case, w and p). So, we have $[(\triangle x)^2/c^2] - (\triangle t)^2 = -16$. This is the first equation connecting $\triangle x$ and $\triangle t$.

To obtain a second equation, we consider the events r and w. According to person A, the spatial distance between r and w is just $\triangle x$ (the same as the spatial distance between w and p, since, according to A, r and p are at the same spatial position, namely right in front of A's nose). On the other hand, according to A the elapsed time between events r and w is $9 - \triangle t$ sec (since A feels that r occurred 9 seconds before p, and that w occurred $\triangle t$ seconds before p; whence A feels that r occurred $9 - \triangle t$ seconds before w). But events r and w are lightlike related, since they are joined by a light-pulse. Hence, their apparent spatial distance must be just c times their apparent elapsed time, that is, we must have $\triangle x = c(9 - \triangle t)$. This, then, is our second equation connecting $\triangle x$ and $\triangle t$.

We now have two equations, $[(\triangle x)^2/c^2] - (\triangle t)^2 = -16$ and $\triangle x = c(9 - \triangle t)$, in the two unknowns, $\triangle x$ and $\triangle t$. We wish to solve these equations. To this end, we substitute the $\triangle x$ given by the second equation into the first equation, to obtain $[c(9 - \triangle t)]^2/c^2] - (\triangle t)^2 = -16$. Cancelling out the cs in the first term, $(9 - \triangle t)^2 - (\triangle t)^2 = -16$. Expanding the first term, $81 - 18\triangle t + (\triangle t)^2 - (\triangle t)^2 = -16$. Cancelling out the $(\triangle t)^2$s, $81 - 18\triangle t = -16$. Rearranging, $97 = 18\triangle t$. So, we have, finally, $\triangle t = 97/18$ sec. To obtain $\triangle x$, we may use the second equation above, $\triangle x = c(9 - \triangle t) = c(9 - 97/18) = c\,65/18$. This, then, is the solution of these two equations.

So we now have that, according to A, the spatial distance between w and p is $c\,65/18$, and the elapsed time between w and p is $97/18$. So, according to A, the speed of person C is given by $[c\,65/18]/[97/18] = 65/97\,c$. That is to say, person A feels that person C is going by at a speed of about 2×10^{10} cm/sec, or about 67% of the speed of light. This is the desired solution.

This result may at first seem a little strange. If A sees B moving at 38% of the speed of light, and B sees C moving at 38% of the speed of light, shouldn't A see C moving at 76% of the speed of light? How could it be that we obtain only 67%? This reaction to our answer to the problem is, I would claim, essentially a holdover from the Galilean view. To see why, let us first briefly recall the standard argument for additivity of speeds. Let us suppose, then, that A sees B moving at some speed v, and that B sees C moving at some speed v'; we want to know how fast A should see C moving. Let 1 second elapse. Then during this second B will move a total distance v ($= v$ cm/sec $\times$ 1 sec) with respect to A, while C will move a total distance v' with respect to B. So, the total distance that C moved with respect to A, during this second, would be the distance that C moved with respect to B plus the distance that B moved with respect to A, or $v + v'$. That is, C moves distance $v + v'$ with respect to A in 1 second. Therefore, the speed of C, according to A, should be $v + v'$. This, then, is the argument that speeds are additive. Let us look at it in a little more detail. Consider, for example, the sentence "Let 1 second elapse." Now, we did not specify for whom the second is to elapse, and in fact we later let this second elapse for two different people (A and B). In short, "Let 1 second elapse" really meant "Let 1 *universal* second elapse." Implicit in this statement was the idea of universality of elapsed time. Consider, as a second example, the distance v' (a number of centimeters) that C moved with respect to B. Later, this distance was simply added in (to the distance that B moved with respect to A) to obtain the total distance that C moved with respect to A. But in its first appearance this distance was a number of centimeters according to B, while in its later use the distance became a number of centimeters according to A. In short, this distance v' played the role of a kind of "universal distance," common to all. Implicit in this part of the argument, then, was the idea of universality of distance. We now know, however, that these spatial and temporal relations are not "universal" in relativity—that only the interval has this honor. It should not be surprising, then, that the everyday argument gives an answer at variance with that from relativity. Again we remark that, for the case when all speeds are much less than the speed of light, the relativity result becomes very nearly addition of the speeds.

(Again, this is seen by redoing the calculation above in the general case.) In this sense, then, relativity "predicts" the everyday observation that speeds are additive.

We also remark that one might even have guessed that in relativity speeds near that of light would not be additive. Suppose they had been additive. Then one could combine a speed of, say, ¾ c with another of ¾ c to obtain a total speed of 1½ c—faster than light. Thus, if one wishes to prohibit particles from being seen going faster than light, one is going to have to do something about additivity of speeds. Relativity cleverly takes care of this problem for us: Speeds are not additive therein, and in fact the total speed is less than the naive sum. Matters work out (as one can see explicitly by doing the general case) so that nobody ever sees anybody else going faster than light.

This completes our calculations of various spatial and temporal observations in the space-time of relativity. A few general points should be noted.

The first point is that the calculations above—comparing apparent elapsed times, apparent lengths, apparent speeds—represent merely illustrative examples. There is, clearly, an infinite variety of conceivable experimental situations—with many different observers passing each other in various ways, with a couple of poles, with everybody taking his own time readings, with light-pulses all over. We have here merely selected a few examples from this vast collection of possibilities. Our examples were not even selected on the basis of any consistent criterion, aside from the general one of simplicity in both the geometrical calculations and the physical interpretations. I more or less made these examples up, and, if asked to produce a set again, might come up with completely different ones. In short, one should not try to "learn" these examples so much as the broader framework they represent.

The second point concerns the general method one uses in attacking such problems. The pattern in every case is the same. One first elicits a detailed statement of the actual physical experiment to be performed, complete with the measurements to be taken. One then represents that experiment by a space-time diagram. Next, one selects numerical values or general letters for certain of the measured quantities, and, by determining intervals and using the fact that intervals are the same for everybody, computes other measured quantities in terms of the given ones. Finally, one interprets those "other measured quantities" physically. The calculations themselves represent a curious mix of physics and mathematics. On the one hand, one must constantly keep in mind what is going on physically. This physical background provides a kind of guide to what one should compute next, what should be compared with what, where the additional light-pulses should be drawn, and so on. Without a clear physical picture, it would be almost impossible to do the calculations. On the other hand, the pictures and calculations themselves are reminiscent of what one sees and does in plane geometry. Indeed, the analogy is perhaps rather closer than mere appearances. Figure 73,*A* illustrates the Pythagorean theorem for a right tri-

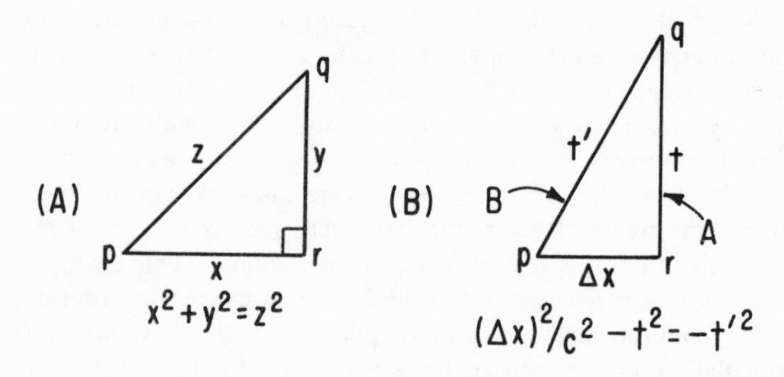

Fig. 73
A comparison between the Pythagorean theorem (*A*) and the formula for the interval in terms of distances and times (*B*).

angle. Figure 73,*B* is a space-time diagram (showing only a part of *A*'s world-line and *B*'s world-line). In this figure, t and t' are elapsed times as seen by *A* and *B*, while Δx is a spatial distance as seen by *A*. The formulas under the figures equate the interval between p and q as seen by *A* to that interval as seen by *B*. We see that it is nearly the same as the Pythagorean theorem, except that in space-time we must introduce c (to convert centimeters to seconds), and we change a few signs. (Orthogonality between rp and rq in the plane geometry example is replaced by A's belief that rp is a "displacement in time" with no spatial displacement, while rq is a "displacement in space," with no temporal displacement.) In essence, then, the calculations in space-time are analogous to calculations in geometry using the Pythagorean theorem. (Drawing in the various light-pulses corresponds to "dropping perpendiculars," and so on in geometry.) Indeed, this analogy often provides a strong hint in doing the calculations. One thinks of the problem as one in geometry, decides what he would do there, and then does the same thing in space-time. One must use this analogy a bit gingerly, combining its insights with those which come from the physics.

The third point concerns those things one has to be careful about in doing such problems. There are two issues on which it is easy to go wrong. First, it is tempting to be sloppy in the statement of the experiment. One might suppose that A "just knows" (by osmosis?) the length of something or the time of something, or what is happening somewhere else. One must take care that nobody is permitted to know anything unless the information actually reaches his world-line by some clearly stated physical mechanism. Second, it is all too easy to slip in Galilean ideas without knowing it. For example, one might implicitly assume that some "elapsed time," or some "speed" will be the same for everybody, a Galilean idea. Usually, these two types of errors are made together. The "just knowing" represents an illegal, Galilean-type transmission of information from one person to another. Learning to automatically be careful about these things just takes a little practice. We might also remark that both of these "errors" correspond to "too much physics and too little geometry." In these problems, the physics acts as the guide, to generate ideas, while the geometry acts more as the censor, to throw out most of the physical ideas, focusing one's attention on the "correct" line.

The fourth and final point concerns the role of the interval in all this. Note that for none of our problems was the final answer of the form "The interval between these two events is . . .

sec^2." Rather, typical answers were "This or that individual sees elapsed time or spatial distance such-and-such." The interval, then, always appears in the middle—never at the beginning or the end. (It is like "the square on the hypotenuse" in the Pythagorean theorem—always converted in the end to "the length of the hypotenuse.") The interval, then, is computed by various individuals in terms of the actual physical experiences they have. Equating intervals then leads to relationships between "real" physical measurements. The interval is a sort of misty thing that stands in the background, dominating all that goes on (for it is the only link we have between the observations of different individuals), but disappearing in the end. It's a beautiful idea—and a rather subtle one—having the crucial quantity be something which doesn't by itself have all that much physical significance. Rather, its significance comes from its implications in these and other similar calculations. (In a sense, a similar thing happens in quantum mechanics, where there the misty thing is the wave function.) One "understands" the interval only through its physical implications. But what more could one ask for as a constituent of "understanding"?

7

Einstein's Equation: The Final Theory

We have now nearly completed the formulation of Einstein's general theory of relativity. All the essential ideas are before us. What we must now do is arrange all these ideas properly, and add one further ingredient: a certain equation called Einstein's equation. In this chapter, we shall first summarize and slightly rearrange what we already have, then add the equation, and finally summarize the theory as a whole. (Why should such rearrangements be necessary? Did we do it the wrong way earlier? The point is that the various ideas were organized earlier in such a way as to provide motivation and to provide more or less direct physical interpretations as we went along. We are now, however, much better able to tolerate a situation in which the motivation and physical interpretations come at the end, where they normally belong.)

We begin, then, by summarizing what we have done already, but now putting it more in the form of a physical theory.

The starting point, now, is the two books—one listing all possible events in the world and the other listing, for every pair of nearby events, the interval between them. Now earlier these two books were introduced as a sort of summary of what our world is like; they were constructed by having people wander around in the world making measurements. Here, however, we wish to take a somewhat different attitude toward these books. Imagine a factory which constantly turns out two-volume sets—one after another, all different from each other. The first book in each set lists a collection of points (labeling them, say, by letters: "point p," "point q," and so on). The second book lists pairs of points, and, for each pair, a number of seconds squared. The total output of this factory, over a long period of time, will be a large pile of two-volume sets. Each of these sets will be called a space-time geometry (note the indefinite article). At present, we are making no commitment as to what relationship, if any, a particular two-volume set—a space-time geometry—is to have with our world, with any "other" world, or with anything physical. They are just two-volume sets of books with names.

Let us now, more or less at random, select one of the sets from the pile (say, that titled *Space-Time Geometry No. 2,831*), and let us hand that set of books to our friend in the closet. He, of course, will not be told that this is merely a small portion of the total output from our factory. Rather, he will probably think that these books were actually compiled, after an enormous experimental effort, by those living in the world outside. Our friend in the closet, knowing relativity as he does, will now begin cranking out a vast number of physical predictions. He will probably start with the light-pulses. "This world-line (He calls it a world-line because he thinks he is speaking of the world outside; he actually expresses it as just some collection of points from his first book.) is a possible world-line for a light-pulse." "This one is not." (Recall how he actually makes these decisions. He can decide from his second book whether or not a given world-line, expressed as a collection of points from the first book, is lightlike. It is these lightlike world-lines that he claims to be possible world-lines for light-pulses.) Then, he would move on to clocks. "This world-line is a possible world-line for a clock. This one is not." (Now, he decides, from the second book, whether or not the line is timelike.) "The elapsed time, according to the clock on this world-line, between events p and q on this world-line (He calls the points events, because he still thinks he is speaking of the world outside.) is 27 seconds." (Recall that such predictions can indeed be made by the person in the closet, from only the two books.) Having now "predicted" all about light-pulses and clocks, he might move on to more complicated things. "The two individuals, *A* and *B*, described by the world-lines illustrated in figure 63 meet at event *p*. Let the events *r* and *s* on *A*'s world-line be so chosen that *A*'s elapsed times are 4 seconds and 9 seconds, as shown. Then *B*'s elapsed time from *q* to *p* will be 6 seconds." "If *A* then finds the event *u* on his world-line which he regards as simultaneous with *q*, then *A*'s elapsed time from *u* to *p* will be 6½ seconds." "Individuals *A* and *B*, and a pole, could be arranged as in figure 71. If *A* now measures 'the length' of the pole by method . . . (he explains it in detail), while *B* measures 'the length' by method . . . , then *A* will obtain length 4.6×10^{10} cm, while *B* obtains length 5×10^{10} cm." The point is that the person in the closet can do all this using only his two books. He first gets the light-pulses and clocks. He already has all the intervals in his second book, and he knows how to interpret those intervals in terms of individuals' physical experiences. He thus goes on and on, through all the examples we have done, through the many we have not done, through example after example. In every case—for every "physical experiment" he can think of—he comes up with definite "physical predictions about what he thinks is the world outside. He proceeds, then, to exhaust all spatial and temporal experiences of (in his opinion, at least) those who live in the world outside. Being a relativist, he is no doubt supremely confident that all his predictions are being checked by those outside, and, to their amazement, are actually being verified.

But, of course, this was all just a trick: What we gave him was just *Space-Time Geometry No. 2,831* by "The Factory"—it had nothing whatever to do with any "world outside." The point of all this is that a given space-time geometry leads to a complete, exhaustive list of detailed, numerical predictions of all experiences of a spatial or temporal character. In other words, a space-time geometry represents a complete model for a physical world. What our factory is doing, then, is turning out a large number of such models. Of course, since these models are being produced by a factory, there is no guarantee that the physical predictions which flow from any one model will actually agree with the physical phenomena we see in our own world. Indeed, since there are clearly many possible space-time geometries that the factory can produce, but only one world, we would expect that most of them will not agree. We are, however, not yet concerned with agreement with our world, but rather, to summarize, with making two points: (1) There exists a vast number of space-time geometries (that is, of two-volume sets), and (2) From each space-time geometry there follows an exhaustive, detailed list of predictions; that is, each space-time geometry represents a model of a physical world.

These preliminary remarks out of the way, we may now state what the theory is as we have developed it so far. Quite generally, any physical theory seems to consist of three things: (1) the statement of a certain class of models; (2) a collection of techniques by which, from a given model, one can make detailed physical predictions about the world; and (3) the following sentence: "All the physical predictions which flow via (2) from one particular model in the class (1) in fact agree completely with the actual physical experiences we have in our world." (If pressed, one might wish to back down somewhat from the overstated, overgeneralized previous sentence.) In terms of this general framework, at least, we can interpret statements commonly made about physical theories. The "beauty" or "aesthetic appeal" of a theory usually refers primarily to how simple the statement in (1) is. There may also be some connotations as to how "natural-looking" (2) is. If the techniques in (2) are not clear (as is surprisingly often the case), one normally tries to clarify them, or, failing that, discards the theory. The "technical simplicity" of a theory more often refers to how simple and straightforward the techniques in (2) are. The "generality" of a theory refers to how large the variety of physical predictions in (2) is. As time goes on, one sees more and more physical phenomena in the world. Insofar as the sentence in (3) remains true—that is, insofar as one continues to be able to find a model in the theory which correctly predicts these phenomena—one's confidence is strengthened. This is particularly so if the range of phenomena one sees is increased dramatically (for example, by acquiring some new experimental apparatus). If one can more or less convince himself that he has seen in our world phenomena that cannot be predicted via (2) from any model in (1), then one regards the theory as being inappropriate for such phenomena. ("Wrong," although perhaps closer to the mark, seems a little severe.) It should perhaps be mentioned that this entire paragraph expresses matters with a sharpness which one seldom sees in practice. There is always some confusion, looseness, and freedom in (2), and often, usually to a lesser extent, in (1); experiments are practically always shrouded in doubts of one sort or another. The business of theory making and theory selecting is very much an art. On the one hand, one must try to make careful, balanced judgments in light of various confusions which always arise in both theories and experiments. On the other hand, one must try to decide how he will weight "beauty," "technical simplicity," "generality," and so on in rating various theories. For the latter, I suppose, one adopts some kind of personal prejudice as to how nature weights these various factors.

It should now be clear how the theory, as we have developed it so far, fits into the general framework of the previous paragraph. (1) The models are the space-time geometries, that is, the output from our factory. (2) The techniques for making predictions from a given model encompass practically everything we have done so far. This includes the representation of physical phenomena in terms of collections of events, the physical interpretations of various time-readings, and so on, all the translations between various geometrical constructs in space-time and physical experiences, and so on and so on. Here are the things the person in the closet would calculate and say; here is the crux of the theory. (3) All the physical predictions which flow via (2) from one particular model in class (1) in fact agree completely with the actual physical experiences we have in our world.

This, then, is the theory as we have developed it so far. As far as I am aware, there is no clear-cut observation (ignoring here quantum mechanics, whose status in all this is perhaps not completely clear) which is at odds with the assertion in (3) above. So, we have here a theory, not a bad one by any means. Why not just stop here? Unfortunately, there are a number of unpleasant features of the theory we have so far. Among these unpleasant features are the following three.

There are more models in class (1) than one would like. As a general statement about theories, the fewer the models in (1), the sharper the theory. The reason is that the actual assertion of the theory is (3)—that there is *some* model that does the job. The more models there are, then, the weaker is assertion (3). If, to take an extreme example, practically anything were admitted as a model, then (3) would be nearly without content, since one could find in one's class some model for almost anything. For our present theory, the models are just being cranked out by a factory, and that is more models than one would like to admit. (I can only assert this from experience with this and other theories. I cannot give a precise criterion for how many models is "too many." One can certainly believe, however, that the vast number of models in the present case is going to seriously erode the predictive power of the theory.)

The second unpleasant feature is that the various physical predictions in (2), while rather complete as regards things spatial or temporal, are seriously incomplete for dynamical phenomena. We will give an example. Fix event p on the world-line of individual A. Individual A decides that, at event p, he will throw off a particle with a certain speed in a certain direction. In terms of our space-time diagram, this decision determines the direction of the world-line of the particle as it emerges from event p. (More precisely, the choice of the particle's direction and speed determines the direction of the tangent to the particle's world-line.) Now we let A issue the further instructions that the particle, once released, is not to be interfered with. It is to fall freely, or move freely, however it wishes. Now we know physically that A's choice of the initial direction and speed, together with A's instructions about leaving the particle alone, will determine uniquely how the particle will move. This is just an everyday physical observation: If one emits a particle, he can determine the event of emission, and the initial direction and speed. Thereafter, if the particle is left alone, one has no additional influence over what the particle does. Thus, the particle's world-line should be uniquely determined by what we have done. Now, we certainly know how to represent this circumstance in terms of space-time: One simply does the experiment, finds the events directly experienced by the particle, and plots those events in space-time. This procedure, however, is inadequate from our new point of view. We are not interested in just representing what happens; we also want to predict what happens. This we cannot do with what we have so far. To put it another way, the person in the closet will be at a loss to answer, from only his books, the question "For this particular experiment, what is the actual world-line of the particle?" (He is not allowed, here, to just give us the various possibilities —the timelike curves through p. We want to know the actual curve.) This, then, is an example of a dynamical phenomenon for which our present theory makes no predictions. (Note, however, that there are some dynamical predictions, for example, if the above experiment were done with a light-pulse replacing the particle.)

The third unpleasant feature is that our theory at this stage seems to have nothing to do with gravitation. We claimed in the introduction that the general theory of relativity was to be, among other things, a theory of gravitation. We have not, however, mentioned the word *gravitation* since. (We note in passing that this third unpleasant feature is rather different in character from the other two. First, it could as much be regarded as an unpleasant feature of the introduction as of the theory itself. Second, as we shall see in a moment, this unpleasant feature is really just a restatement of the first two.)

So, we are now in possession of a theory of sorts, and of three objections to it. The natural impulse would be to try to take care of the three objections one at a time, that is, to first try to cut down on the number of models, then to try to introduce some reasonable dynamical law for how particles move, and so on. It turns out, remarkably enough, that all three can be taken care of in a single stroke.

The "single stroke" is a certain equation called Einstein's equation. We may write it symbolically as follows:

$$\begin{pmatrix}\text{Curvature of}\\ \text{space-time}\\ \text{geometry}\end{pmatrix} = G \begin{pmatrix}\text{Mass density}\\ \text{of matter}\\ \text{in space-time}\end{pmatrix}$$

(Why only symbolically? Up until this point, we have generally been true to the mathematics which underlies the theory. Although a more mathematical treatment would have helped to clear up a few confusing notions—such as that of "nearby"—nothing essential in this mathematics has been suppressed. Here, for the first time, we are not able to continue in this spirit. Einstein's equation itself, written out properly and fully, requires either a branch of mathematics called differential geometry or else a rather lengthy excursion into some technical geometrical constructions. It is, for our purposes, not worth the time that would be required to carry out either program. We shall therefore have to be content with a symbolic statement of the equation, together with a discussion of what it means and does.) We proceed, then, to an explanation of what the three things appearing in this symbolic equation mean.

We begin with the term "curvature of space-time geometry." First consider the following question: How does one know that the surface of the earth is curved? An obvious answer is that one can go up in airplanes or satellites, and then actually see and photograph the curvature of the earth. Alternatively, during an eclipse of the moon one can see the shadow of the earth on the moon, and one notes the curvature of that shadow. Suppose, however, that we had agreed to disallow methods which involve having ourselves or other instruments outside or above the surface of the earth; we want the measurement to be done entirely within the surface itself. That is to say, for example, we might imagine the earth inhabited by small, very flat, two-dimensional ants, which crawl about the surface. These ants have no opportunity to move off the surface, nor to see or be influenced by anything off the surface. We ask, Would such ants be able to determine whether or not the surface of the earth is curved? In order to answer this question, we might begin with a more general question: What is the sum total of all the information (of a geometrical character) that these ants could accumulate about the surface of the earth? They could certainly locate various points on the surface of the earth, and they could certainly measure distances between pairs of points (by crawling between the points)—and that is about it. In short, the ants can acquire precisely that information about the earth which is already contained in our two books for the earth. One book, recall, lists all possible points on the earth, the other the distance between any two nearby points on the earth. (Remember, also, that from this information one can determine the distance, once the route is specified, between more distant points.) We ask again, Can the ants determine (or, what is the same thing, can we determine from only the two books) whether or not the surface of the earth is curved?

The answer, it turns out, is yes. Consider the triangle on the earth in figure 74. One vertex is at the North Pole, the other two on the equator. Now, the ants could certainly draw such a triangle. Furthermore, they could determine each of the angles of the triangle. (This could be done, for example, as follows. Con-

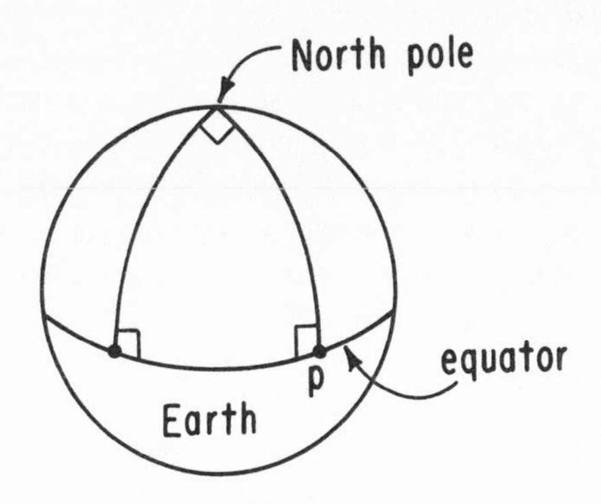

Fig. 74
A "triangle" on the surface of the earth. The "sides" are a portion of the equator, and two longitudes. All three angles are right angles.

sider the vertex p, and complete a small triangle near that vertex as shown in figure 75. Now measure the lengths of all three sides of this small triangle. Finally, use, for example, the law of cosines from trigonometry to compute the angle θ.) We have ar-

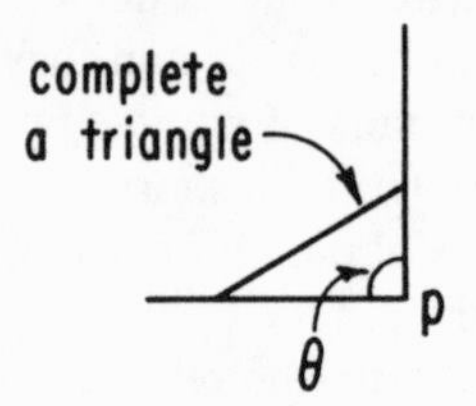

Fig. 75
Measurement of an angle in the plane using only distances.

ranged figure 74 so that all three angles of the triangle are right angles: That the two angles on the equator are such is clear, while we have adjusted the angle at the North Pole to be a right angle, too. Since the ants can measure the angles, they will be aware of this fact. The ants now consult their Euclidean geometry books, and there learn that the sum of the angles of a triangle (in *plane* geometry) is 180°—that, in particular, no triangle may have all its angles right angles. Thus, by combining what they read in their geometry books with what they actually measure on the earth, our ants can conclude that they are certainly not living on a plane. (Those familiar with spherical trigonometry will recall that the spherical excess of a triangle on a sphere is defined as the sum of its angles minus 180°, that this spherical excess is always proportional to the area of the triangle, and that the proportionality constant is essentially the square of the radius of the sphere. Thus the ants in effect "measure" the radius of the earth—without ever leaving the surface!) Our conclusion, then, is that it is indeed possible for the ants to determine that the surface of the earth is curved, that is, that the information contained in the two books for the earth suffices to detect the presence of curvature.

The surface of the earth is like space-time; the ants are like us. The ants cannot leave the surface of the earth (by dictum, to make the analogy); we certainly cannot leave space-time (because of the way space-time was introduced). The ants have access only to points on the earth and distances between nearby points; we have access only to events in space-time and intervals between nearby points. The ants, from the information they have, can detect curvature; we, from the information we have, can detect curvature. This analogy suggests, then, that, using only geometrical constructions involving the events in space-time and the intervals, we may introduce a quantity which may be thought of as the "curvature of space-time geometry" itself. This "curvature of the space-time geometry" is determined mathematically from the events and intervals (that is, from the information in our two books) by a certain geometrical construction not unlike that described above for the ants on the surface of the earth. (The geometrical construction itself is, however, considerably more complicated for space-time than for the earth. On the one hand, one must deal with four dimensions [in space-time] rather than two [for the earth]. Furthermore, the simple and familiar Pythagorean theorem [which leads, for example, to the cosine law for triangles] for the case of the earth must be replaced consistently by our interval-formulas, with the resulting sign changes and use of c.) This analogy, then, is just an analogy. It merely gives one an overall picture of how the construction might work, while hiding numerous technical details. It is here (that is, use of an analogy rather than a detailed mathematical definition of "curvature of space-time geometry") that we are suppressing mathematics. In any case, such a definition exists, and the result is called the "curvature of the space-time geometry." It is this quantity which appears on the left side of Einstein's equation.

The use of the word *curvature* above is perhaps a bit misleading (the word is used here, of course, only because it is traditional). In the case of the earth, one thinks of the "real curvature" as that which one actually sees from above the earth. The thing that the ants calculate from within the surface seems more like a bit of mathematical trickery which just happens to give the same answer. In the case of the curvature of the space-time geometry, however, we only do it by the "mathematical trickery" method. What about the "real curvature"? Is space-time "really curved"? The best answer I can give is this. Consider again the curvature of the earth. In that case we have two equivalent methods of determining this curvature. Has one any genuine grounds for preferring one over the other? As far as I can see, one does not. The only reason for preferring the former (from above the earth)—for regarding it as "more real"—is that it is perhaps more familiar. In the case of space-time, however, we have available but a single method for finding a "curvature-like quantity"—the intrinsic method. (The other method is, of course, unavailable, since we cannot "leave space-time.") Since for the earth one has two equivalent methods, since their common result is called the curvature of the earth, and since only one of these methods has an analogue for space-time, it is perhaps natural to call the resulting quantity "curvature of the space-time geometry." In any case, one should think of this quantity as simply the mathematical result of a certain geometrical construction (analogous to that described above for the earth) in terms of the events and intervals for space-time.

Before leaving the subject of curvature, we make one further observation. Consider again the curvature of the earth. Now in fact, there are many different regions of the earth at which this curvature could be measured. If, for example, the curvature were determined near the (rounded) top of a mountain, one would obtain a rather large value; if measured over the surface of an ocean, a somewhat smaller value. The curvature of the earth, then, is not just a single number, but rather depends on where one is on the earth. The curvature of the earth, in short, varies over the surface of the earth. A similar remark applies to the curvature of the space-time geometry. It is not a single number; rather, it varies over space-time. One region of space-time, then, might have a rather large curvature, another a smaller curvature, and some other region may have zero curvature. We shall see shortly how this remark ties in with Einstein's equation.

We are, recall, in the process of explaining what the various quantities which appear in Einstein's equation mean. We have now completed the discussion of the left side. The first quantity appearing on the right, G, stands for Newton's gravitational constant, $G = 6.7 \times 10^{-8}$ cm^3/gm-sec^2.

We come, finally, to the other quantity appearing on the right, "mass density of space-time." This quantity is essentially what its descriptive name suggests. It expresses how much matter there is, in grams per cubic centimeter, in any given region of space-time. Thus, for example, in a region of space-time in which there is no matter at all, this quantity would be zero; in a region in which there is a great deal of highly compressed matter, this quantity would be rather large. We note that the mass density of space-time is also not a single number; it, too, varies over space-time.

Einstein's equation, then, requires that the curvature of the space-time geometry be equal to the constant G times the mass density of matter in space-time. The equation is to be applied at each event in space-time, that is, it is really many equations, one for each event. In more detail, what this means is as follows. Fix an event p in space-time. Then one can compute (from the two books which describe the space-time geometry) the curvature of space-time around p. Further, one can measure the mass density of matter around p. Einstein's equation, for this event p, then requires that the former equals G times the latter. The full equation thus requires that this equality hold for *every* event p in space-time.

Let us see where we are. At the beginning of this chapter, we introduced a certain physical theory (in which the models were arbitrary space-time geometries). This theory was found to suffer from three difficulties: admitting too many models, failing to account properly for dynamics, and failing to account for gravitation. We then claimed that all three difficulties could be taken care of in one stroke. We then wrote down Einstein's equation, and explained what it means. Clearly, our next task is to show how Einstein's equation is able to handle these three difficulties.

We consider first the problem of cutting down the number of models in our earlier theory. Let us return now to that enormous pile of two-volume sets (each representing a "space-time geometry") that our factory has turned out. Let us select one of the two-volume sets from the pile at random. Let us think for a moment of this space-time geometry (a mere work of fiction, of course) as representing some actual physical world (as the person in the closet would), and let us in particular regard this world as endowed with matter of various types. We may now do the following calculation. On the one hand, we may compute, from the two books, the curvature of the space-time geometry, say at some fixed event p. The result will be just some value for the curvature at p. On the other hand, we may measure the mass density of the matter in space-time at the event p. The result will be some value for the mass density at event p. Having determined these two things, we may ask ourselves, Will it necessarily be true that Einstein's equation is satisfied at event p? Obviously, the answer is no. We just made up the space-time geometry (or rather, our factory did), and there is not the slightest reason to believe that the curvature that is computed from it at p should equal anything in particular. Of course, it could happen by coincidence that the curvature of this particular space-time geometry at p just happens to equal the value we have chosen for the mass density at p, but this would be only a coincidence.

Our conclusion then is that, given any space-time geometry and any distribution of matter, Einstein's equation will not automatically be satisfied. (Indeed, we had better hope not. An equation in physics which is automatically satisfied doesn't say anything, and therefore is usually not worth writing down.) The imposition of Einstein's equation, then, will serve to cut down the number of models.

In fact, there are a number of different facets of this general idea, which correspond to different shades of meaning for Einstein's equation. It is perhaps worthwhile spending a little time exploring them. To this end, let us now build, at great expense, a new factory. This new factory will produce three-volume sets of books. The three books in each set will be called *The Events*, *The Intervals*, and *The Matter Distribution*. The first volume will simply list the events, as before. The second volume will list, for every pair of nearby events, some value for an interval between them, as before. The third volume, however, is new. It will list, for every event which appears in the first book, a value for the "mass density in space-time" at that event. A typical listing in the third book, then, would read "event p-mass density. . . ." Once again, the books are written more or less at random. Again, the total output will be a great pile of three-volume sets.

Let us now select, from this pile, one of the three-volume sets. We open the first book, and select some event, say p, from it. We now open the second book, which has all the intervals in it. From this second book, we carry out the mathematical construction (which we have here described only by analogy) which yields finally the value of the curvature of that space-time geometry at event p. We write the answer down on a piece of paper: "Curvature at $p = \ldots$" Finally, we open up the third book, and find the listing which gives the value of the mass density at p. We also write this down: "Mass density at $p = \ldots$" Now we may "check" Einstein's equation (at event p). We simply find out whether or not what we have written down for the curvature at p equals G times what we have written down for the mass density at p. Obviously (since the books have been produced at random) equality will in general not hold. If it does happen to hold (a mere coincidence), we move on to some other event, say q, in the first book, and continue the process. Continuing in this way, we do the same thing for every single event in the first book. Now, there are two possibilities: It could happen that we find Einstein's equation to hold for every event in the first book, or it could happen that we manage to find some events at which it does not hold. In the case of the first possibility we write, on the outside of the box holding the three books, "does satisfy Einstein's equation" (at every event); in the case of the second, "does not satisfy Einstein's equation." In this way, we may separate the output from this factory into two piles, consisting of those sets which do or do not satisfy Einstein's equation. Again, Einstein's equation serves to cut down on the number of models, but now it is doing so for a different type of "model" (namely, a three-volume set, rather than a two-volume set).

Now, however, we are in a position to look at Einstein's equation from other vantage points. Suppose, for example, that you were sent one of these three-volume sets, in its box, through the mails. The parcel, however, arrives damaged. The first two books are in good condition, but the third is completely unreadable. On the box, however, you can still make out the words "does satisfy Einstein's equation." It is unnecessary, in this situation, to go through a long correspondence with the factory to get this matter straightened out. One can actually reconstruct the third book from what one has already received. The procedure, of course, is straightforward. One takes a blank book. Look up in the first book some event, say *p*. Then compute, from the second book, the value of the curvature of the space-time geometry at *p*. Finally, dividing this computed curvature by *G*, one obtains what must have been the value of "mass density of the matter" at *p* in the third book (since this three-volume set satisfies Einstein's equation). Thus, one writes in one's blank book "event *p*-mass density. . . ." Continuing in this way, through all the other events in the first book, one completely restores the third book.

What we are saying, then, is that Einstein's equation can be regarded as determining the mass density of matter from the curvature of the space-time geometry. From this viewpoint, then, the equation doesn't really restrict the space-time geometry in any particular way. Rather, it provides the possibility of making additional physical predictions from that geometry alone, predictions as to the mass density of the matter. From this viewpoint, then, Einstein's equation represents an additional technique by which physical phenomena (the phenomena of the mass density of matter) are determined from the space-time geometry.

There is still another viewpoint. Suppose that the second book had been damaged in transit (but the box still said "does satisfy . . ."). One now has possession only of the list of events and the mass density of each event. From this, and the knowledge that Einstein's equation was satisfied, one can compute, without actually having the book with the intervals, what the curvature of space-time geometry was. This, it turns out, is a bit less information than was originally contained in the second book—one obtains only the curvature (which, of course, would normally be computed from the intervals) but not the intervals themselves. From this point of view, then, Einstein's equation is regarded as determining the curvature of the space-time geometry from the (given, from this viewpoint) mass density of matter. That is, Einstein's equation is now giving "geometrical, interval-like information" from physical information about the mass density. Here then is also a sense in which Einstein's equation "restricts the space-time geometry."

For this last viewpoint, it turns out, there is a good physical analogy. Consider a tightly stretched rubber sheet, held horizontally above the earth's surface by a wooden frame. The rubber sheet would be flat. Think of the rubber sheet as representing the "geometry of space-time" and in particular its curvature (zero, right now) as representing the curvature of the space-time geometry. We next introduce matter as follows. Take a small steel ball, and place it on the sheet (fig. 76). The ball will

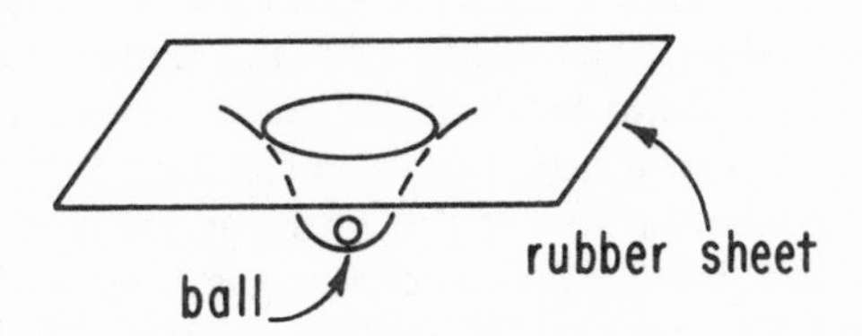

Fig. 76
An analogy to Einstein's equation. Matter, placed on the rubber sheet, causes curvature of that sheet.

then "sink into" the sheet, stretching it and in particular causing it to bend. Our rubber sheet now displays curvature (curvature which, we note, could also be computed by ants living on the surface of the sheet), where this curvature could be regarded as being "caused" by the placing of the steel ball on the sheet. More generally, if some more complicated distribution of matter were placed on the sheet, then the sheet would bend (be "curved") in some more complicated way. In this physical analogy, then, a distribution of matter (here, placed on the sheet) causes curvature (of the rubber sheet). The situation is analogous to that of the above interpretation of Einstein's equation, in which a distribution of matter in space-time causes curvature of space-time.

As in all analogies, this one has its strong and its weak points. An obvious strong point is that it involves the same words ("matter causes curvature") that one might apply to Einstein's equation. There is, in fact, a second and much deeper strong point. It turns out that the analogy is a rather deep one mathematically. That is to say, the mathematical equations which would describe the bending of the sheet as a result of matter placed thereon are very nearly the same as the mathematical equation we call Einstein's equation. (Again, however, there are differences involving the dimension, signs, and the appearance of c.) In any case, many of the salient features of Einstein's equation (at least, those which arise from the present viewpoint toward that equation: "matter causes curvature of space-time geometry") have direct analogues for matter placed on a rubber sheet. There is, in addition, one serious weak point. In the case of a matter distribution on the rubber sheet, the fact that the matter causes curvature in the sheet is not at all a "fundamental law of physics." Rather, it is a consequence of other, more fundamental, things: the gravitational attraction of the earth for the steel balls, the equations which describe how a rubber sheet bends and stretches when subjected to such forces, and so on. In the case of Einstein's equation, however, the situation is quite different. One does not, at least at the present time, regard Einstein's equation as being a consequence of other, more fundamental things in physics. One does not think of the matter as being "pulled against the space-time geometry" (by some external mechanism); one does not think of the space-time geometry as "reacting to this pulling by adopting a curved configuration." Rather, one thinks of Einstein's equation as being the fundamental thing all by itself. The equation asserts by fiat that a distribution of matter requires a certain curvature in the space-time geometry.

This last interpretation of Einstein's equation—"matter causes curvature in the space-time geometry"—is perhaps the most important one. Nonetheless, the interplay between the various interpretations—the making of transitions at the appropriate point between one interpretation and another—represents a significant aspect of the subject.

We come now to the second of the three difficulties with our earlier theory: the absence of definitive dynamical predictions in that theory. How is Einstein's equation going to help us here? It does so in a subtle and elegant way. Recall that we allowed individual A at event p to emit a particle which was free to move as it wished thereafter. The problem was that we knew that the particle would describe some particular, unique world-line under our physical instructions. On the other hand, one could draw numerous world-lines starting at p in the given direction. The theory, then, seemed unable to even begin to predict what the particle would do after it was released. Now, however, we are in possession of Einstein's equation, and this changes the situation completely. To fix ideas, let us fix the space-time geometry once and for all, and let us consider a number of possibilities for how the emitted particle might move. A few such possibilities are illustrated in figure 77. Now, for each one of these five possibilities, we may construct one of our three-volume sets. The first two books of each set will always be the same, for we are in every case dealing with the same events and the same in-

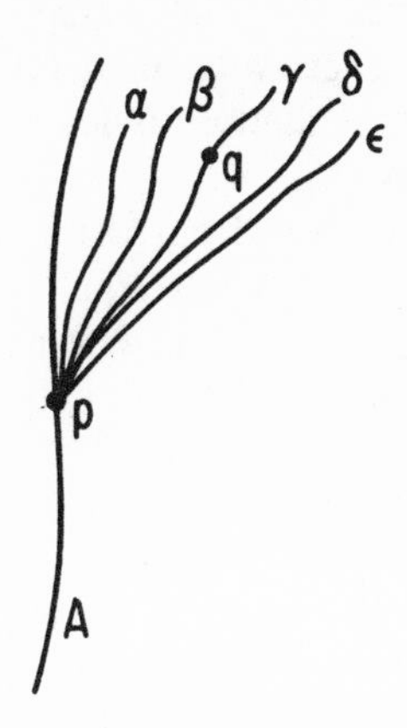

Fig. 77
Given event p, and the initial direction of a world-line from p, there are many world-lines having that direction at p. Which of all these possibilities will actually be realized by a thrown particle?

tervals. (This is what we meant by "fix the space-time geometry.") The third book in each of the five sets will, however, be different. A particle, after all, has mass, and so its introduction into space-time represents some mass density. Consider, for example, event q. The sets to represent situations α, β, δ, and ϵ will all say, in their third books, "event q-mass density zero," for in these situations there is no particle experiencing event q. The set representing situation γ, however, will say, in its third book, "event q-mass density . . . (not zero)," for in situation γ there is a particle experiencing event q. In short, each of these possibilities represents one of our three-volume sets. But we have seen that the three-volume sets can all be divided into two piles, those which do and those which do not satisfy Einstein's equation. It is clear that not all five of our sets will deserve to be labeled "does satisfy Einstein's equation"; at most one will be so labeled, and, more likely, none will be so labeled. (It's not easy to get a "does satisfy Einstein's equation" by blind luck.) In short, we must be very careful indeed about where we are to represent the particle as going if we are to expect to end up with something which satisfies Einstein's equation. In this way, then, Einstein's equation restricts what matter is allowed to do. We thus see that Einstein's equation at least offers the possibility for some dynamical laws.

The discussion above was from the standpoint of one particular interpretation of Einstein's equation, that in which the space-time geometry is regarded as determining the distribution of matter (and therefore, in particular, determining how particles must move). We may also see the same thing from the other interpretation, in which matter causes curvature and thereby influences the space-time geometry. From this standpoint, one would think of all the space-time events as having been given a priori, and the matter distribution as having been given also. From this information we may already determine the world-line of the emitted particle (for we merely look up in the third book which events have a nonzero mass density associated with them, and collect all such events together to obtain the world-line of the particle). What we do not yet have, however, is the information of what the intervals are. We now impose Einstein's equation. It says here that these intervals cannot be chosen arbitrarily, but rather must be so chosen that the resulting curvature of the resulting space-time geometry satisfies Einstein's equation, that is, is equal to the (known, in this case) right side of that equation. In this case, then, Einstein's equation no longer acts to restrict the actual world-line of the particle. Rather, it acts to restrict the intervals, given that world-line. What, now, is required for *A* to make physical statements about what the particle is doing (for example, "the particle seems to be speeding up and rushing away right now")? Two things, of course, are required: the world-line of the particle and also the intervals. (The intervals, note, also play a role, for these intervals determine how clocks tick, how light-pulses move, and so on. But these are the very instruments which *A* will use to make his measurements on the particle.) From this interpretation, then, Einstein's equation works as follows. It cannot do anything about the world-line, but it cleverly adjusts what the intervals are to be so that *A* ends up making the "correct" (that is, physically realized) statements about what the particle is doing. In this way, Einstein's equation again provides dynamical information about how the particle moves in space-time (or, more precisely, information about what *A* will have to say about its motion).

From either interpretation we obtain the same result. On imposing Einstein's equation, one loses his former right to specify both the space-time geometry *and* the world-line of the particle (or, what is the same thing, the mass density associated with the particle) at will. Rather, these two things must be connected via Einstein's equation. The result of this connection is that the dynamical behavior of physical phenomena is now tied down to space-time itself.

Again, we may make a physical analogy. Consider again our rubber sheet, and let it be in some curved configuration such as that shown in figure 78. (This configuration could have been created, for example, by some distribution of matter placed on the sheet. We are not interested in this matter right now.) Let us now place on this curved sheet one of our small steel balls. The

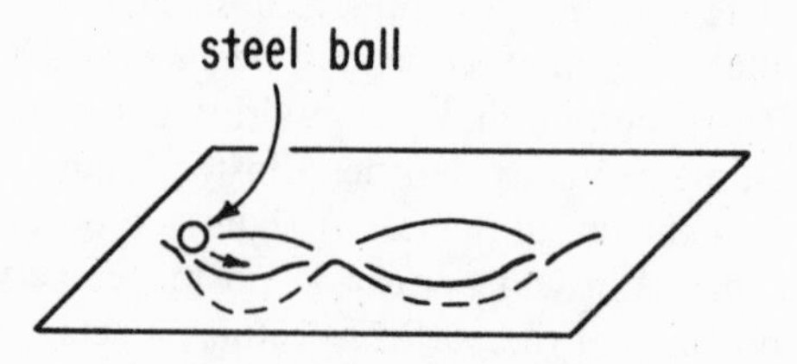

Fig. 78
An analogy to Einstein's equation. Curvature (of the rubber sheet) causes dynamical effects of matter.

ball will of course "roll down and around the various bumps on the rubber sheet." The result is that the curvature of the sheet dictates in detail how the matter (in this case, the steel ball) will move. This is in essence what happens in relativity with Einstein's equation. Matter in the world is forced, via that equation, to react to the curvature of space-time in one way or another. This "reaction to the curvature" is what, within the theory, dictates how matter moves about.

The present analogy again has the strong point that it involves the same words ("curvature affects matter") as those associated with the situation in space-time. Mathematically, the analogy is weaker than our earlier one. The weakness of the earlier analogy persists in the present one. Thus, the rolling of the ball on the rubber sheet is understood in terms of other, more fundamental, things (the attraction of the earth for the ball, and the slope of the surface of the sheet). The effect of curvature on matter in relativity, via Einstein's equation, is regarded as fundamental by itself. Finally, there is an additional weakness associated with the two analogies taken together. In the case of the rubber sheet, there are really two separate mechanisms acting. One is that responsible for the curvature of the sheet under the matter distribution (earth attracts matter distribution, attractive forces bend sheet); the other is that responsible for the motion of the ball in response to the curvature (earth attracts steel ball, ball rolls under influence of this due to slope of rubber sheet). In the case of relativity, on the other hand, there are not so many disparate things coming into play. It is all done by means of a single equation acting in and interpreted in various ways. From the conceptual point of view at least, the situation in space-time is actually simpler than that of the physical analogy.

We turn, finally, to the last of our three difficulties: What does all this have to do with gravitation? This is easy. We have already seen above that Einstein's equation can be interpreted as requiring that "matter cause curvature in space-time," and that it can also be interpreted as requiring that "matter move in certain ways in response to curvature in space-time." Gravitation thus arises as follows. Let there be, say, two massive bodies in the world. Then, according to Einstein's equation, each will cause a certain amount of curvature in space-time. Further, according to Einstein's equation, each body will be forced to move in a certain way in response to the curvature caused by the other. The net result, then, is that each body influences the other. It is just this influence that we call gravitation. The idea is that the curvature of the space-time geometry acts as a sort of intermediary between the two bodies. It is essentially the gravitational field.

We may at this point push our analogy still further (thereby weakening it still further) as follows. Let there be placed, on our rubber sheet, two steel balls, as shown in figure 79. What will

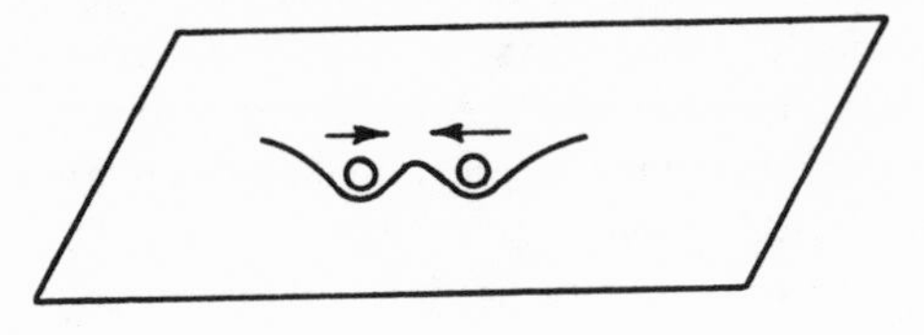

Fig. 79
The analogies of figure 76 and 78 taken together: There is a "force" between any two objects in the universe.

happen? Intuitively, it is clear that the two balls will roll together at the center of the sheet. One could say, if one likes, that some sort of force causes the two steel balls to attract each other, and that they come together as a result of this attraction. One could also say it as follows: Each ball causes curvature in the rubber sheet. Each ball reacts to the curvature of the other ball. The result is that, under this intermediary of curvature, the two balls come together. Now, the first version above ("The balls just attract each other") is analogous to Newton's law of gravitation. There, no "mechanism" for this attractive gravitational force is provided; one simply asserts its existence. The second version is analogous to relativity. One "explains" gravitation as an effect due to the causing and reaction to curvature in the space-time geometry. (Of course, one does not "explain" Einstein's equation. But the basic constituents of a physical theory are never "explained"; they just sit there asserting things, waiting for the next theory.)

Now, at long last, we are prepared to state what the final theory is. First, the models. The models consist of our three-volume sets (that is, of a space-time geometry together with a distribution of matter) which satisfy Einstein's equation. (That is to say, not every possible three-volume set is now admitted as a model.) The physical interpretation of a model is of course the same as it has always been, namely essentially everything which has appeared in these pages so far. All of these ideas, techniques, computations, descriptions, interpretive remarks, and so on constitute the framework in which the mathematical and geometrical content of a model is translated into physical notions. The third part is always the same: the assertion that the physical interpretations associated with one of our models correspond to the various physical phenomena we see about us in our world. We note that gravitation now comes into the theory automatically, because of the restriction on the models (they must satisfy Einstein's equation). This, then, is the general theory of relativity.

Having now stated what the theory is, we will proceed to make a few comments about it.

One might very well be left with the impression that the theory itself is rather hollow: What are the postulates of the theory? Where are the demonstrations that all else follows from these postulates? Where is the theory proven? On what grounds, if any, should one believe the theory? I can only answer these questions with my own opinions. It seems to me that "theories of physics" have, in the main, gotten a terrible press. The view has somehow come to be rampant that such theories are precise, highly logical, ultimately "proved." In my opinion, at least, this is simply not the case—not the case for general relativity and not the case for any other theory in physics. First, theories, in my view, consist of an enormous number of ideas, arguments, hunches, vague feelings, value judgments, and so on, all arranged in a maze. These various ingredients are connected in a complicated way. It is this entire body of material that is "the theory." One's mental picture of the theory is this nebulous mass taken as a whole. In presenting the theory, however, one can hardly attempt to present a "nebulous mass taken as a whole." One is thus forced to rearrange it so that it is linear, consisting of one point after another, each connected in some more or less direct way with its predecessor. What is supposed to happen is that one who learns the theory, presented in this linear way, then proceeds to form his own "nebulous mass taken as a whole." The points are all rearranged, numerous new connections between those points are introduced, hunches and vague feelings come into play, and so on. In one's own approach to the theory, one normally makes no attempt to isolate a few of these points to be called "postulates." One makes no attempt to derive the rest of the theory from postulates. (What, indeed, could it mean to "derive" something about the physical world?) One makes no attempt to "prove" the theory, or any part of it. (I don't even know what a "proof" could mean in this context. I wouldn't recognize a "proof of a physical theory" if I saw one.)

If the above is really the case—if one is not concerned with proving the theory—what, then, do relativists do with their time? This is a difficult question to answer, because the activities are so diverse. One major area of activity could be summarized under "trying to figure out which model, if any, corresponds to our physical world as we see it." This broad title, however, covers a lot of ground. One might, for example, take some specific system, such as a galaxy or a neutron star, find some model (or, more often, approximate model) which seems to correspond to that system in a general sense, and then, by looking at the more detailed physical phenomena associated with that model, try to make further predictions about how the system should behave. In this way, one makes astrophysical predictions from the theory. Another area of activity works from the opposite direction. One begins with some model one has happened to find by some method or other. One works out a variety of physical predictions associated with it. One then asks (or looks) to see if this particular constellation of predictions seems to be actually realized anywhere in our world. Of course, there is no guarantee that any model one happens to come up with will turn out to be realized in our world; it is all very much hit and miss. Another broad area of activity consists of finding new techniques, or refining the old, by which physical interpretations can be obtained from models. Here, one is not normally concerned with any specific model, but rather with a broader question: Given any model, how can one more completely, more clearly, or more definitively extract its physical content? A third area of activity consists of looking for new theories. Here, one is in effect going through the same process we have just gone through, but making various (possibly major) changes along the way. There are clearly numerous possibilities along these lines.

One might object to the theory on the following grounds. Looking at the region of space-time in our immediate vicinity, we do not see any obvious sense in which it is curved. How, then, can the theory presume to assert that space-time is curved? The answer to this objection is that it is all a question of "how curved." Consider again the surface of the earth. We know that that surface is curved, but nonetheless one would have little chance of detecting it if one spent one's entire life in Chicago. Although there is indeed curvature, it does not manifest itself significantly over small regions of the earth. To see the curvature clearly, one must travel over a larger region of the earth's surface, or, at least, receive information from a larger region. The situation in space-time is exactly the same. If we look only in our local region of space-time (say, in one's laboratory, or in the earth and its immediate environment) one does not allow oneself sufficient room to see clearly the effects of the curvature of space-time. To see significant effects of the curvature, one must examine much larger regions of space-time. We do in fact see some local effects of the curvature, for example, dropped stones fall to the earth, but the clear, dramatic, geometrical effects require larger regions. It should at this point be clear why the primary testing ground for general relativity turns out to be astrophysics. In this field, one is concerned with distant stars, galaxies, and so on, that is, one is concerned with information received over rather larger regions of space-time. It is in this field, then, that one would have the best chance of actually directly observing curvature of space-time.

Finally, we add one remark concerning how our earlier development (up to the beginning of this chapter) fits into the framework described in this chapter. Earlier, one notes, we did not at all take the present point of view. We did not consider a variety of models, nor did we try to decide which of these corresponds to our own world. Rather, what we did was try to proceed directly from the physics of our world to some particular model. This was, it can now be revealed, essentially a technique of presentation. It would have been confusing to start right at the beginning with the models, with no idea how a given model is to be interpreted physically. What we had to do, then, was to get the interpretation out before treating the question of models at all. It would have been equally confusing to try to "interpret" in some general sense, without specific reference to the particular world in which we live. What we did, therefore, was build up the interpretative techniques and a particular model (namely, ours) both at the same time, as a device to introduce the interpretation at an early stage. Once these interpretative techniques are available, however, one is free to change the viewpoint as we have done. One is free to consider models all by themselves in the abstract, and one is free to interpret them physically (what the world "would" be like with that model) despite the fact that any given model may not correspond very closely with our own world. Having made this change in viewpoint, the central question becomes our central question: Which of the many available models corresponds most closely to the world as we see it?

An Example: Black Holes

Black hole is the label that has been assigned to a certain model (more precisely, to a certain class of models) within the theory. That is to say, one of our three-volume sets—one that satisfies Einstein's equation—has *black hole* written on its box. This model was actually discovered, historically, in a search for solutions of Einstein's equation. Since we have written Einstein's equation only symbolically, we cannot go through the historical process by which this particular space-time was discovered. We can, however, treat the far more interesting and important aspect of this subject, namely, the statement of what this space-time is and what its physical implications are. This is what we propose to do in the present chapter. One crucial point should be understood at the beginning. For the bulk of this chapter, we shall be dealing with the question of what this particular space-time is and what the physical phenomena are to which it leads. One must remember, however, that this is just one of the three-volume sets in our pile. We do not suggest, initially, that these phenomena are actually realized in our world. We are not, at the beginning, dealing with our world at all. Rather, we are simply going through the motions for some particular (but rather interesting) space-time. At the end, we shall make a few remarks on the question of whether or not this particular model corresponds in any sense to our world.

Our first task is to say what the space-time is, that is, to express the contents of the first book (listing the events) and the second book (listing the intervals). Clearly, it is going to be hopeless to reproduce these two books in their entireties. We must find some reasonably concise way of expressing the essence of what is printed there. This we do as follows. First the events. For the set of all possible events in this space-time, we draw ordinary three-dimensional space. (We are not, of course, suggesting that these "events" are "in" ordinary three-dimensional space, only that ordinary space provides a medium to draw a picture of the set of events.) In this picture we draw a vertical cylinder and its axis, as shown in figure 80. (The cylinder and axis are, for the moment, just benchmarks to help us locate things and describe what is going on geometrically. Any physical interpretations attached to them will have to come later.) We now call different regions of this picture by names, as follows: The region outside the cylinder is called the external region, the cylinder itself is called the horizon, the region inside the cylinder (excluding the axis) is called the internal region, and the axis itself is called the singularity. We emphasize that these names, for the moment, serve only to make it easier to describe verbally where a point

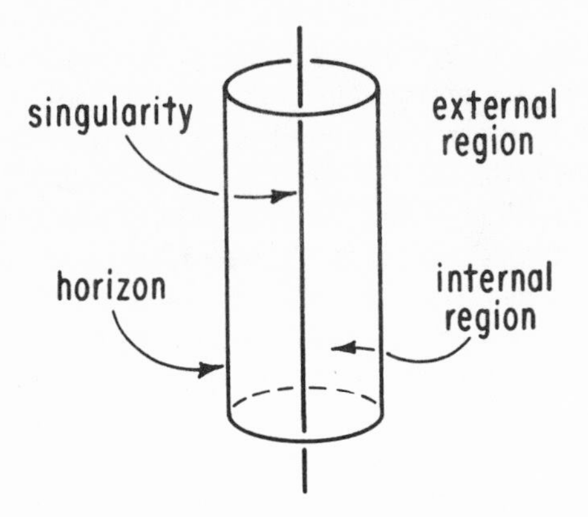

Fig. 80
The set of events for a black-hole space-time. We divide the events into three classes: those in the external region, on the horizon, and in the internal region.

or something else is in the diagram. Of course these names also have physical and geometrical connotations. Such connotations are to be ignored for the moment, and will be justified eventually.

The picture above, then, summarizes for us the contents of the first book. We must next express the contents of the second book. Clearly, we are faced with a new problem, for we can hardly list every single pair of nearby points in the diagram above with, for each pair, an interval between them. We must therefore find some way to summarize the essential content of the second book without writing out the actual listings. This we do as follows. Fix some point p in this space-time. Our second book will then list, for every point q near p, the interval between p and q. From this information, we may determine the light-cone of point p, namely by drawing in space-time the locus of points q which have zero interval from p. Thus, from the intervals we may determine the light-cone of point p. But these light-cones are rather simple geometrical objects—they may easily be drawn in our space-time diagram. What we shall do, therefore, is just "attach" to each point in our diagram a small cone which will represent the light-cone of that point, as determined from our second book.

This attaching of light-cones certainly conveys within our diagram some of the information in the second book. It should be emphasized, however, that not all of the information in the second book has in this way been brought into the diagram. We ask, What can one tell about the intervals, having in one's possession only the light-cones? First of all, we can certainly tell whether two nearby events have zero interval, for if q has zero interval from p, that is, is lightlike related to p, then either q will lie on p's light-cone or p will lie on q's light-cone. We may also determine whether or not points p and q are timelike related, for in this case q would lie inside p's light-cone or else p would lie inside q's. Finally, we may determine whether or not two events are spacelike related (neither on nor in the light-cone of the other). In short, what we can determine, with just the light-cones, is precisely the sign of the interval between two nearby events: whether they are spacelike, timelike, or lightlike related.

The interval itself of course gives more information than just its sign. This additional information seems rather difficult to represent in our diagram, so for present purposes we must leave it out. The result will be that our physical interpretation of the resulting space-time geometry will be limited to a certain extent, but, as it turns out, this limitation is not at all severe.

So, what we must now do is attach, to each point in figure 80, a small light-cone, in this way bringing some of the information from the second book into the figure. The resulting light-cones are those shown in figure 81. (Note that we are not at this point claiming to justify our choices. Rather, we are simply summarizing the contents of two books labeled *black hole.*) In the figure, the light-cones in the external region far from the horizon look rather ordinary (for light-cones): They sit up straight. As one moves in toward the horizon, however, the light-cones in the figure begin to do two things: they become narrower and they begin to tip in toward the horizon. This "narrowing and tipping" continues all the way up to the horizon. The light-cones of points on the horizon itself are so tipped that the cones are tangent to the cylinder. (More precisely, the vertical direction in the diagram from any point on the horizon is both tangent to the cylinder and tangent to the light-cone of that point.) As one proceeds into the internal region, the light-cones continue to become narrower, and continue to tip. As one approaches the singularity, finally, the cones have become very narrow indeed, and have tipped to about 45° in the diagram. In figure 81, we have drawn only the light-cones for a few representative points. What are the light-cones of the other points? We have omitted these just to avoid cluttering the picture, but they can be recovered as follows. First, the entire picture is to remain the same under vertical translations up and down. Thus, if one wants to know the light-cone of some event such as q in the picture, it will be the same as the light-cones of all points vertically above or below q in the picture. In particular, it will be the same as the light-cone of the point p in the picture. From this remark, then, it suffices to know all the light-cones of all points on one horizontal plane in the diagram, since, by vertical translations, one can then obtain all

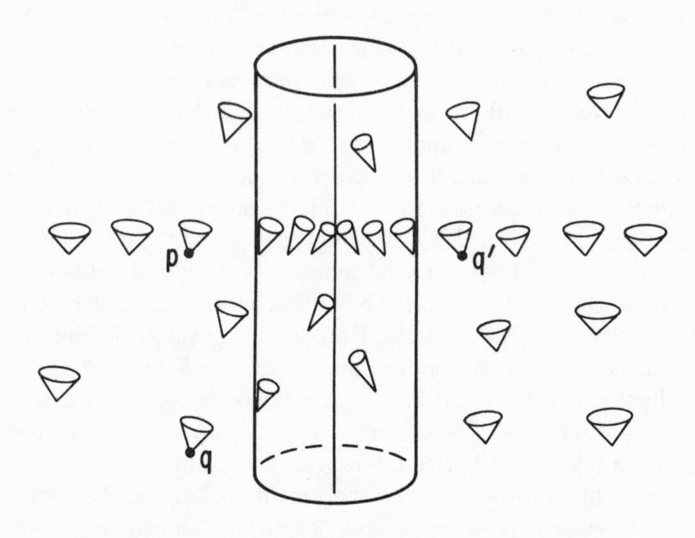

Fig. 81
The light-cones for the black-hole space-time. The cones are "vertical" in the distant external region, become tangent to the horizon, and then, in the internal region, lean toward the singularity. It is intended that the figure have the symmetries of translation up and down, and rotation about the axis.

the light-cones of all points. Second, the entire picture is to remain the same under rotations of the diagram about the axis. Thus, if one wants to know the light-cone of some event such as q' in the picture, for example, one can note that a rotation about the axis carries point p to point q'. Acting on the light-cone of p under this rotation, one obtains the light-cone of q', as shown. From this remark, it follows that if one knows the light-cones of all points on a horizontal line from far away to the axis, then one knows already the light-cones of all points on the horizontal plane containing that line (for, under rotations about the axis, the line generates the plane). For one particular horizontal line reaching the axis (namely, that coming into the picture from the left), we have described the light-cones in some detail. Using rotation, we can thus obtain all the light-cones in that horizontal plane. Using vertical translation, we can, finally, obtain all the light-cones for all the points in the picture. To get a little practice with the diagram, one should try a few rotations and vertical translations, to verify that various assorted light-cones we have drawn in the diagram are correct.

This, then, is our space-time geometry (rather, a part of it: we have given the points, and the signs, at least, of the intervals). Two consolidating remarks are in order.

The first remark concerns the points on the axis. These are certainly points of the diagram itself, that is, they are points of the Euclidean three-dimensional space in terms of which the space-time is represented. They are not, however, to be regarded as real events in this particular space-time geometry. (There is nothing deep happening here; such points are just omitted from the first book.) The fact is that such points would have to be omitted. As one approaches a point on the axis, such as the point p shown in figure 82, the light-cones from all directions are becoming very narrow, and are tipping toward the axis. If p is to be a point of our space-time, then p must have a light-cone.

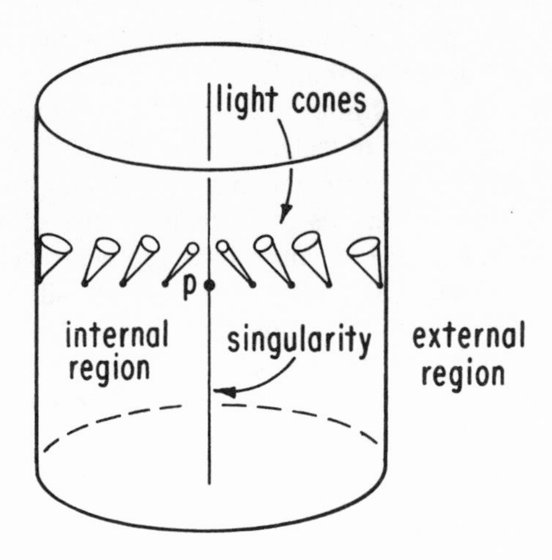

Fig. 82
More detailed view of the light-cones near the singularity.

What, however, is the light-cone of p going to be? It is clear that there is no reasonable choice, for the light-cones, as one approaches p, approach different things depending on the direction of approach to p. Thus, since there is no consistent choice for a light-cone at p, we must just omit such points from our space-time geometry. It should now be clear why the axis is called the "singularity." As one approaches the axis, the light-cones have "singular behavior," with the consequences that one is forced to omit the axis-points from our space-time geometry.

The second remark concerns suppression of dimensions. We are here drawing a three-dimensional picture of a four-dimensional space-time. What happens in the suppressed dimension? The space-time diagram we have drawn is axially symmetric. The actual four-dimensional space-time labeled *black hole* is spherically symmetric. What we have done, then, is suppressed one of the spatial directions. (It makes no difference which one, since it is spherically symmetric.) This diagram, then, is rather similar to the space-time diagrams in figure 14 representing the earth. In that case, the spherically symmetric earth became a solid cylinder in an axially symmetric space-time diagram. One should keep in mind, as we discuss the diagram, that there is an extra, hidden, dimension around.

The road before us is now clear. On the one hand, we have available a great variety of techniques by which physical predictions can be made to flow from a given space-time geometry. On the other, we have in our possession (the essence of, at least) a space-time geometry. All we must do is apply our techniques to this particular space-time geometry. The result will be a multitude of physical predictions about black holes. This point cannot be overemphasized. Having selected our space-time geometry, all our manipulative freedom in this problem has ended. We cannot later interject new facts about black holes at whim. All we are now permitted to do is routinely work out various physical consequences and gaze at those consequences. (Of course, this particular model may not correspond to our world, so we still have an escape clause if we become too unhappy with those consequences.)

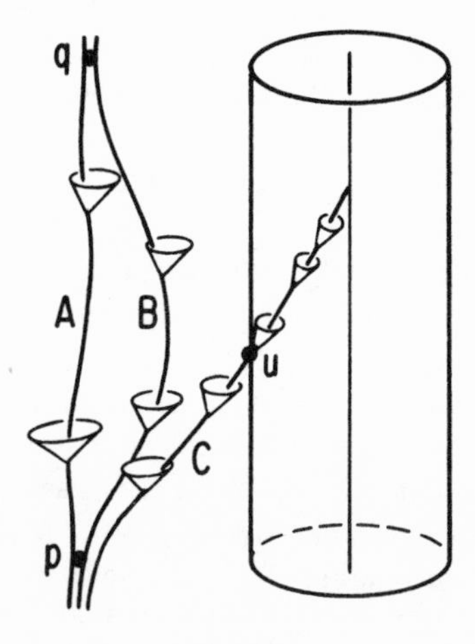

Fig. 83
Three observers in our black-hole space-time. Observer *A* remains in the distant external region, *B* ventures somewhat closer to the hole, and *C* passes through the horizon to the interior of the hole.

We begin, then, with applications of the following interpretative technique. The world-lines of individuals are timelike lines. Three typical such world-lines are shown in figure 83. To facilitate the discussion below (and only for this purpose: one should not take this sentence too seriously), one may think of "the black hole" as a kind of "object" whose "world-tube" is the internal region. The world-line *A* represents an individual who keeps well away from the black hole. Were *A* orbiting about the black hole, his world-line would be represented instead by a spiral, as in figure 14,*A*. Individual *B* has somewhat more curiosity about the black hole. He begins with individual *A*. At event *p*, however, he decides to take a closer look for himself. He thus goes in, remaining in the external region, closer to the horizon. Eventually, he becomes tired of this exploration, and goes back out to individual *A*, joining him at event *q*. Individual *C* is overwhelmed by curiosity about this black hole. He goes directly toward it and then, at event *u*, crosses the horizon. He then remains in the internal region for a stretch, and finally hits the singularity (that is, his world-line goes to this region in the diagram). (We note that all three world-lines are indeed timelike; they remain inside the light-cones.)

Recall that there is no event available to *C* on the singularity itself. In particular, there is no possibility for a further extension of *C*'s world-line after it hits the singularity. What, then, does "the world-line of *C* hits the singularity" mean physically? Mathematically, what happens is that his world-line just stops. Physically, this would mean that *C* is "snuffed out of existence"; after some finite time according to himself, he ceases to exist in space-time. From the point of view of *C*, it is easy to say what this means. He exists at his watch-reading of 1, at his watch-reading of 1½, at his watch-reading of 1¾, at his watch-reading of 1⅞, and so on, but he simply does not exist for watch-readings of 2 or greater. There is just never an event at which *C* will say "I am experiencing this event, and I am noting that my watch reads 2" (or anything greater). Is this, one might ask, the same as having *C* "disappear before one's eyes"? Not necessarily. This last question describes, more or less, an actual experiment that might be performed on *C* (namely, one in which there is a second individual, who receives light-pulses from *C*, and who interprets for himself things about *C* in terms of those pulses). In order to predict what will happen for such an experiment we must draw in the diagram these actual light-pulses, and find out, from the diagram, where these pulses will go. We shall turn shortly to such questions of visual appearances.

Let us now suppose that *C*, at some point on his world-line, was handed the space-time diagram for this black hole. By looking at the diagram, he can tell that, if he continues on as he is, he is going to be "snuffed out." Not wishing such a fate, he decides therefore to stifle his curiosity and return to the safety of *A*. Will he be able to do so? It all depends, as one might have expected, on how soon (at which event) *C* makes his decision to try to return to *A*. Three possibilities are shown in figure 84. Suppose first that *C* makes his decision at event *v*. Now, *v* is in the external region. Although the light-cones at *v* have begun to

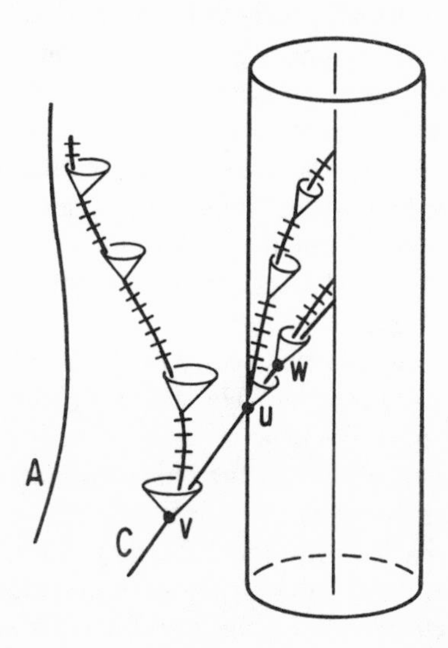

Fig. 84
The result of *C*'s deciding, at various events along his world-line, that he wishes to return to the external region.

tip toward the black hole to a certain extent, the "back side" of the cone at v is not yet vertical (for this first happens at the horizon). Now at event v individual C has the choice (by, say, accelerating with a rocket) of having his world-line go in any direction within the light-cone of event v. Since, however, the light-cone at v has not yet tipped "too far," he can have his world-line begin to go back out toward A. That is, C has the option of having his world-line, beyond v, be the railroad track shown. Thus, C begins to move away from the black hole. As he gets farther and farther away, the light-cones become more and more "straight up." It thus becomes even easier to continue moving away from the hole. He is thus able to return to A. Next, let us suppose that C makes his fateful decision at event u, on the horizon. The light-cone at event u is that shown. Again, C has the power to adjust his world-line, beginning at u, in any way he wishes, provided only that its direction always lies inside the light-cone. But by event u the light-cone has already tipped so much that it is already tangent to the horizon. Any direction from *u* inside that light-cone is going to take one into the internal region—closer to the singularity and farther from *A*. So, one moment later, no matter what he does, *C* is going to be in the internal region. But in the internal region the light-cones are tipped even more toward the singularity. Again, *C* has no choice (since he must stay within the light-cone) but to continue on toward the singularity. The closer he gets to the singularity the more tipped the light-cones become, and the more *C* is forced to head for that singularity. Clearly, if *C* makes his decision at event *u*, it is already too late: He is going to be snuffed out no matter what he does. Event *w* is even worse for *C*. Here, he is already in the internal region, the light-cones are already tipped too far, and he must reach the singularity.

We may summarize as follows. If one is born in the external region, then one always has available the option of remaining forever in the external region. In exercising this option, one still may go close to the black hole, explore, and return to one's friends who have remained farther out. One must, however, be very careful not to cross the horizon. Once the horizon is reached (no matter what one's intentions were), one must proceed to the internal region, and, shortly thereafter, one will reach the singularity and be snuffed out. This form of suicide can be carried out by one in the external region whenever one wants, that is, if one has hesitated for ten years to go into the internal region, one then still has the option of having this experience, for one is never cut off from directing one's world-line across the horizon. In short, "crossing the horizon" can be experienced at most once.

We now consider a slightly different aspect of *C*'s decisions above. Shown in figure 85 is what *C*'s world-line would be if he does nothing. Now, at event *v* individual *C* suddenly remembers that he has a dentist appointment at event *q* in the external region. (Dentist's offices are all in the external region.) At the occurrence of event *v*, however, *C* is in the middle of a particularly exciting conversation or other activity. Individual *C* would therefore like to wait as long as he can before beginning his return journey to event *q*. How long can *C* wait? Let us, for definiteness, say that the elapsed time along *C*'s "does nothing" world-line between *v* and *u* is 10 seconds. Clearly, then, if *C* waits for a total of 10 seconds after *v*, then he will already be at *u*, and then it will be too late to make his dentist appointment (or any other appointment in the external region). Suppose, then, that he waits, say, 9½ seconds after event *v*, that is, until event *s* in the dia-

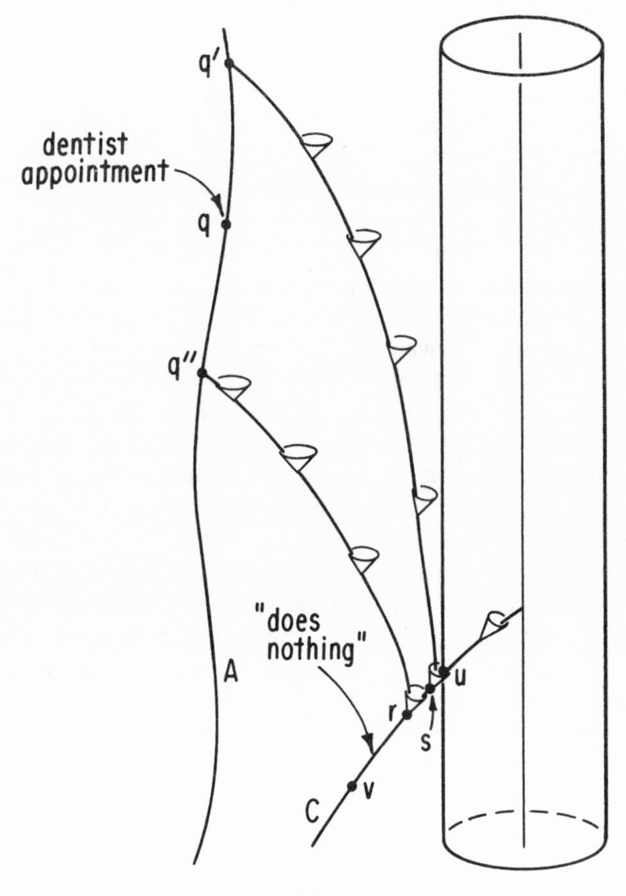

Fig. 85
The nearer to event *u* that *C* arrives at his decision to return to the external region, the later *C* arrives in this region.

gram. At event *s*, then, *C* begins to rush back to *A*. Now *C*, of course, can only direct his world-line within the light-cone of event *s*. Event *s*, however, is so close to the horizon that its light-cone is tipped so far that the back side of that light-cone is very nearly vertical. Thus, beginning at event *s*, individual *C* cannot make much headway toward *A*. He can, however, by accelerating very hard away from the black hole, cause his world-line (which always must remain within the light-cone) to inch away from the horizon. After a bit, *C* is somewhat farther from the hole, whence the light-cones have straightened up somewhat. Now, with the light-cones less "tipped," he can begin to make more headway toward *A*. Unfortunately, however, *C* had to waste all that "vertical motion" in the diagram in order to get away from the black hole. The result is that *C* only makes it back to *A* at event q' as shown. Individual *C* has missed his dentist appointment. Suppose, then, that *C* had begun his return journey 8 seconds after *v*, namely at event *r* shown in the diagram. Now *r* is somewhat farther from the horizon than *s*, and so the light-cones are somewhat less tipped there. The result is that, beginning at *r*, individual *C* makes more headway toward *A* right from the start. In this case, he easily makes it for the appointment (and, indeed, arrives early, at event q'' shown).

The point of all this, then, is the following. Those last few seconds before event *u* are crucial. No matter how far along *A*'s world-line the dentist appointment is (that is, according to *A* at least, no matter how "far into the future"), *C* must be careful not to wait too long if he is to meet the appointment. Of course, *C* can always get back to *A*, no matter how close to *u* he allows himself to be. The point is, however, that the closer *C* gets to *u* before deciding to return to *A*, the later (according to *A*) *C* will eventually return. Individual *C* can even miss an appointment seven years in the future, if he waits, say, 9.9999 seconds after *v* before starting back to *A*.

Let us now consider another situation (fig. 86). Again, we have individual *A* who remains in the external region, and individual *C* who goes through the horizon and gets snuffed out. In this case, however, *C* has a friend, *D*, who happened to be busy

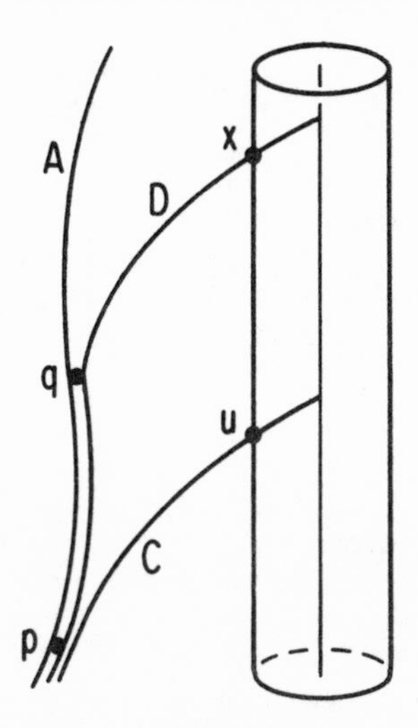

Fig. 86
A space-time diagram representing two individuals, C and D, who enter the black hole, D later than C.

at event p, and so who did not join C on his trip through the horizon. At event q, however, D finishes his business and decides to join his friend. Now D knows of course that C is going to be snuffed out, but he wishes to at least join C in the last few moments before this happens. (Of course, if he is successful in doing this, then D, having entered the internal region, isn't going to survive either.) The question is, Is it in fact possible for D to join C? The answer is clearly no. Beginning at event q, individual D is forced to keep his world-line inside the light-cone. This prevents D, for example, from going "horizontally in" to C. (Physically, such motion would correspond to "moving faster than light.") In keeping his world-line timelike, D must "move upward" in the diagram as he "moves inward" toward the black hole. The result is that D only crosses the horizon, say, at event x in the diagram, continues thereafter to the singularity, and never meets up with C.

Let us now modify the above slightly (fig. 87). We keep A's world-line the same, and D's world-line the same. In this case,

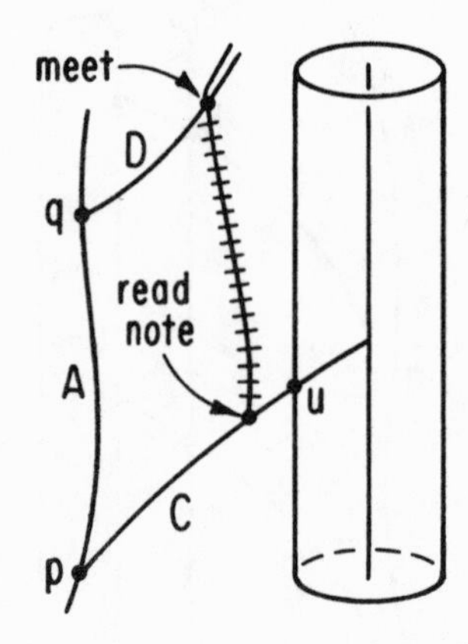

Fig. 87
The result of *C*'s deciding, before event *u*, to return to *D*.

however, *D* hands *C* a note, at event *p*, saying "At event *q*, I shall begin a trip into the hole, in order to be with you during those last moments before you-know-what." Now clearly, if *C* should happen to open and read the note before event *u*, then *C* could arrange matters to actually meet up with *D*. What *C* would do, of course, is to immediately begin accelerating away from the hole, as shown in the diagram. Individual *C* would thereby remain in the external region, and he could simply wait until *D* comes by, at which event they could adjust their world-lines to be together. (Indeed, they would even have the choice, at this event of meeting, of remaining together in the external region or going together through the horizon, and so on.) The more interesting case, however, is that in which *C* doesn't open the note until event *u*, the event of *C*'s crossing the horizon (fig. 88). Having experienced *u*, *C* is of course now committed to entering the internal region and reaching the singularity. We ask, Is it nonetheless possible for *C* to arrange matters, beginning with event *u*, so that he will meet up with *D*? The answer, it turns out, is still yes. At event *u*, the light-cone is so tipped that its back side is vertical. Individual *C*'s only obligation is to keep his world-line within the light-cone. Thus, C can direct his world-line to be as "nearly vertical" as he wishes. Having so chosen this direction at u, C will of course move into the internal region, away from the horizon. As one moves inward, the light-cones tip still further, and thus C is forced to go more and more toward the singularity. However, C nonetheless had the option, at event u, of making his world-line as nearly vertical as he wished. That is to say, C had the option of "gaining as much vertical motion in the diagram as he wishes, during his inching in toward the singularity." Clearly, by accelerating very hard, beginning at u, away from the singularity, C has the option of getting "as far upward in the internal region as he wishes" before his final plunge

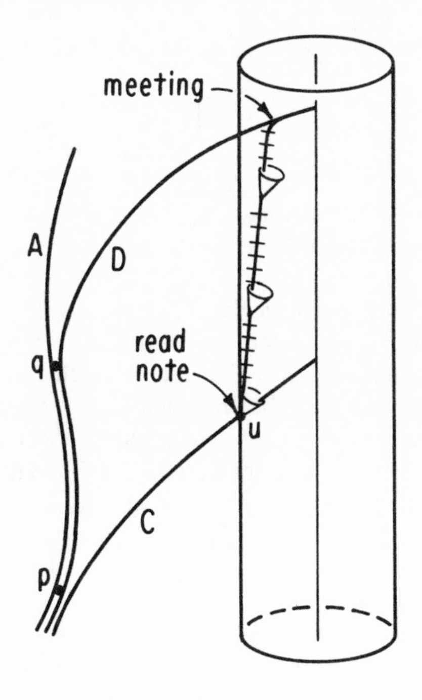

Fig. 88
The result of C's deciding, at event u, to return to D. Now, C and D are only able to meet within the internal region.

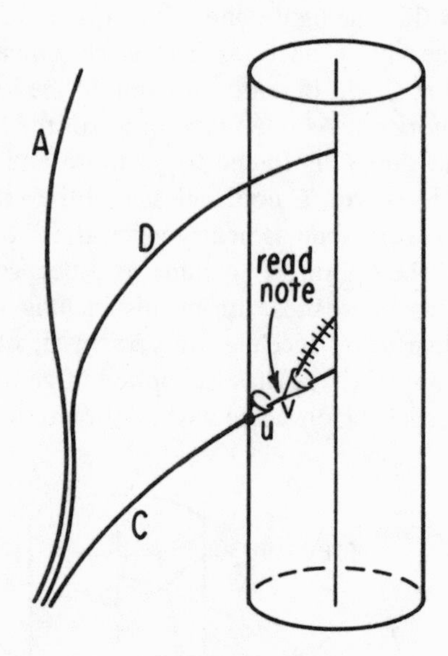

Fig. 89
The result of *C*'s deciding, after event *u*, to return to *D*. In this case, *C* is unable to do so.

into the singularity. In particular, *C* can arrange matters so that he meets up with *D*, as shown in figure 88.

Finally, we might consider the case in which *C* only opens the note somewhat after *u*, as shown in figure 89. Now *C* has a more serious problem. At event *v* (note opened), the light-cones are already tipped past the vertical (since *v* is in the internal region). Now *C* no longer has the option of making his world-line as "nearly vertical as he wishes" (for "too vertical" would take his world-line out of the light-cone of *v*, which is illegal). In short, at *v* individual *C* is already committed to a certain rate of approach to the singularity as he tries to move vertically in the internal region in order to meet *D*. Thus, if *v* is too far along, then *C* will no longer be able to meet up with *D*. Thus, it is possible for *C* to open the note after *u* and nonetheless meet *D*, but it has to happen rather quickly after *u*, for if *C* delays too long, then his option to ever see D again will have been lost.

Recall what we are doing. We have a certain space-time geometry (introduced in fig. 81) which we are trying to interpret physically. This interpretation is carried out by applying to the particular geometry the various interpretative techniques we have at our disposal. So far, we have applied just one interpretive technique: that the world-line of an individual is timelike. From this alone, we obtain all the physical discussion of the last several pages. (Clearly, many more such examples could be done.) We turn now to a second interpretive technique: The world-line of a light-pulse is a lightlike line. This statement clearly allows us to draw, within our space-time diagram, the world-lines of various light-pulses. Since it is through such light-pulses that one "sees" things visually, we are now in a position to discuss what various goings-on will look like visually to various individuals.

We consider first the following situation. Our friend A remains in the external region. Individual A continually emits light-pulses inward toward the black hole (fig. 90). We may draw in the

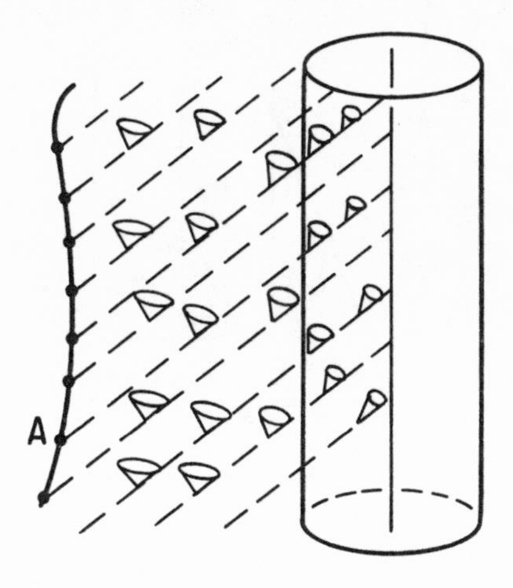

Fig. 90
Space-time diagram representing observer A's emission of light, which enters the black hole.

world-line of each successive pulse using the fact that these world-lines are lightlike (that is, that the world-lines are tangent to the light-cone). The resulting light-pulses have the world-lines shown. Note that each goes right through the horizon, and right on in to the singularity.

The situation above is not a very interesting one, because it is difficult to become too involved about what happens to some light-pulse. Let us now introduce somebody who will "see" (visually) this light from *A*. Let us introduce individual *C*, as shown in figure 91. Events *s*, *t*, *u*, *v*, and *w* have been indicated on *C*'s

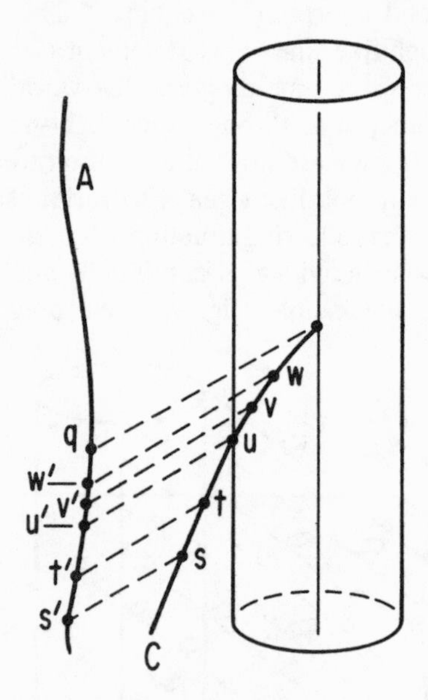

Fig. 91
Light from observer *A* as seen by observer *C*, who enters the black hole. *C* is able to maintain surveillance of *A* during his journey.

world-line, and corresponding events *s′*, *t′*, *u′*, *v′*, and *w′* on *A*'s world-line. The light-pulse emitted by *A* at event *s′* reaches *C* at event *s*; that emitted by *A* at *t′* reaches *C* at *t*; and so on. What, then, will *C* see visually if he looks at *A* during his trip into the hole? At successive events on *C*'s world-line, he will experience the light emitted from successive events on *A*'s world-line. There is nothing particularly strange about this; it is what happens for ordinary people looking at each other in everyday life. Nothing particularly strange happens even as *C* crosses the horizon. He just continues, at successively later events, to experience the light emitted by *A* at *A*'s successively later events. Nothing dramatic even happens as *C* approaches the singularity. Finally, *C* is snuffed out. He never experiences any light emitted by *A* after event *q* in the diagram. (This is a little strange, I suppose, but, after all, what can one expect when he gets snuffed out?) So, our conclusion is that *C*, watching *A*, will just continue watching *A*, and then eventually *C* will be snuffed out.

What would happen if, at event *u*, *C* suddenly became very curious about *A*, and wished to watch more of *A*'s activities? Certainly, *C* could enable himself to do this. What *C* must do is immediately begin, at *u*, to accelerate away from the singularity, causing his world-line to be more like that illustrated in figure 92. Having done this, *C*'s world-line will intercept more of the light-pulses from *A*'s world-line, that is, *C* will be able to see more of *A*'s activities. Clearly, *C* can see as much of *A* as he wishes (depending on how nearly vertical he wishes to make his world-line at *u*), but *C* can never enable himself to watch *A* forever.

Our conclusion so far, then, is that individuals who enter the internal region and reach the singularity can continue to gaze visually at the external region, seeing various goings-on out there. (Of course, these individuals cannot go back out to that region.) Suppose now that *C*, after he has entered the internal region and while he is gazing at the external region, happens to spot somebody he would like to meet. What can *C* do? He certainly cannot go back to the external region, so his only hope would be that this person eventually decides to enter the internal region. Of course, *C* would immediately adjust his world-line to be "as vertical as possible" (consistent with the light-cones) so as to have the greatest chance of meeting this person should that person decide to enter the internal region. It might then dawn on *C*

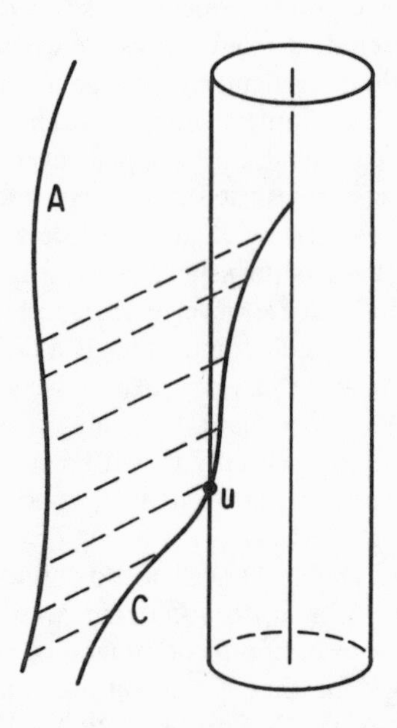

Fig. 92
By accelerating away from the singularity just after event u, C is able to watch A's activities for a longer time.

that he should try to send a note out to that person, informing the person of the possibility of coming in for a meeting in the internal region. It should be clear that this will never work. The note (say, tied to a rock and thrown) is in any case going to describe a timelike world-line. As such, the note will never get out of the internal region (since timelike world-lines don't make it). Thus, there is no way to send such a note to the external region. In short, C can do no more than move "as vertically as possible" in the internal region, hoping that the person just decides to enter the internal region. If not so decided, C is just stuck.

These, then, are some of the visual impressions of an individual going into the black hole. Let us now consider a somewhat different situation, in which the light is emitted by the person, C, going into the hole. Once again, we may draw in the world-lines of the successive light-pulses, as shown in figure 93. The behavior of light-pulses emitted by *C* are more interesting. Those emitted from events, such as *v* in the diagram, in the external region just go out into the external region. The closer the emission point on *C*'s world-line is to the event *u* at which *C* crosses the horizon, the "farther up in the diagram" the light-pulse goes before it eventually gets well away from the hole. (Geometrically, we are here simply drawing in the lightlike lines. The situation is in fact very much like that illustrated in fig. 85. If one begins very near the horizon, then the light-cones are so tipped there that even

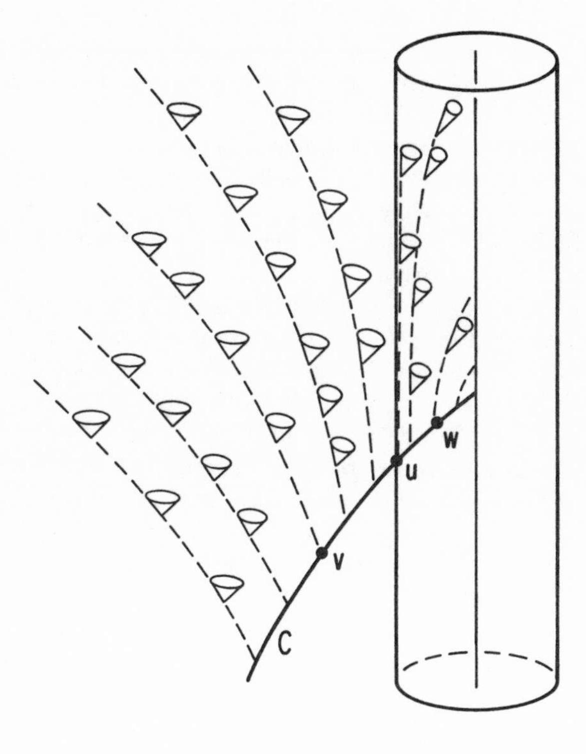

Fig. 93
Light-rays as emitted by an observer *C* who enters the black hole.

light makes very little headway getting out into the farther regions of the external region. The light does make it eventually, but the pulse is forced to move very much in the vertical direction while just inching outward.) What happens to the light-pulse emitted by *C* at event *u*? The world-line of this pulse is just the vertical line tangent to the horizon (a lightlike line, we note). Finally, we have the world-lines of the light-pulses emitted by *C* after event *u*. These pulses just go right into the singularity. The longer (according to *C*) after *u* that the pulse was emitted, the more directly it goes to the singularity. In particular, pulses emitted just after *u* (from, for example, event *w* in the diagram), only work their way to the singularity, meanwhile sweeping vertically for a great amount. (This situation is very much like that illustrated in fig. 88. If one is very near the horizon, in the internal region, then the light-cones have tipped past the vertical, but not by very much. The result is that the emitted light-pulse goes "very nearly vertically" for a stretch. Eventually, however, the world-line works its way nearer the singularity, where the light-cones are tipped further. The result is that the light-pulse eventually bends in toward the singularity.)

In any case, figure 93 shows the world-lines of successive light-pulses emitted by *C*. Now, we are in a position to decide what someone watching *C* visually would see (fig. 94). First, let us permit individual *A* to receive these light-pulses. A succession of events on *C*'s world-line are indicated, together with corresponding events on *A*'s world-line corresponding to *A*'s receipt of the light-pulses from *C*. What, then, does *A* see visually? At successive events on *A*'s world-line, *A* receives the light from successive events on *C*'s world-line. However, no light from *C* after event *u* (on the horizon) ever reaches *A* (for the world-lines of such light-pulses either remain on the horizon or enter the internal region, while *A* remains in the external region). Thus, *A* is only able to see *C* during *C*'s stay in the external region. Does *A* see *C* disappear somehow as *C* reaches *u*? Not really. The point is that, although *A* may watch *C* forever (according to *A*), all *A* will see visually is *C* getting closer and closer to the horizon. That is to say, at some event *A* might receive the

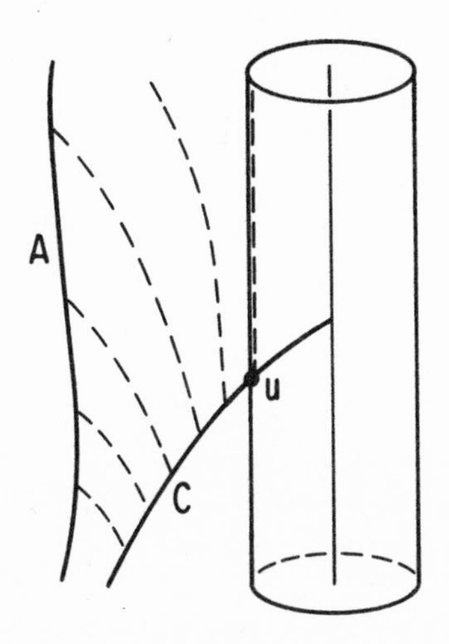

Fig. 94
A watches *C* by means of the light received from *C*. In this case, *A* remains outside the hole, while *C* goes in. The result is that *A*, at any of his moments, can still see *C*, but that *C*'s activities after entering the hole are forever hidden from *A*.

light *C* emitted 1 second before *u* ("1 second" according to *C*, of course). An hour later, *A* will be receiving light *C* emitted, say, 0.3 seconds before *u*. A day later, *A* will be receiving light *C* emitted, say, 0.02 seconds before *u*. And so it goes. Even after hundreds of years, *A* will still be receiving light *C* emitted just before *u* (now, perhaps, 0.0000007 seconds, according to *C*, before event *u*). Individual *A* never visually sees *C* crossing the horizon, never sees *C* in the internal region, and certainly never sees *C* reaching the singularity.

We might note in passing another conclusion from the remarks above. According to *A*, individual *C* seems (that is, when watched visually) to slow down as time goes on. This is clear because it takes *A* forever (according to him) to see *C*'s last 1 second before *C* experiences event *u*. Thus, for *A* years and years seem to pass while *A* only sees visually *C*'s going through some small fraction of a second of *C*'s life. Clearly, this slowing down gets worse and worse (according to *A*) as time goes on. Eventually (say, after a couple of years), *C* will appear to *A* to be effectively frozen.

Next, let us introduce individual *D*, who stays with *A* for a while, watching *C* (fig. 95). At event *q*, however, *D* becomes rather bored with all this (particularly since *C* appears to be

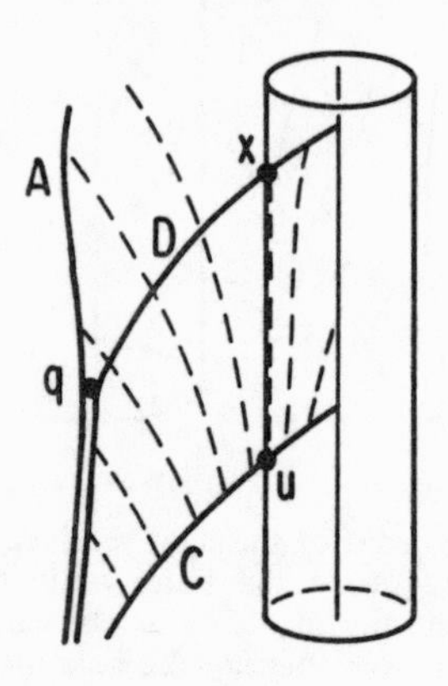

Fig. 95
Observer *D*, by deciding to enter the black hole, is even able to watch *C*'s entry into the black hole.

changing so slowly that nothing much is happening) and decides that he would like to visually see *C* crossing the horizon and entering the internal region. Now, *D* knows that this can never be accomplished if *D* remains in the external region (since light from the event "*C* on the horizon" just goes up the horizon, while light from events on *C*'s world-line in the internal region remains in the internal region). So, *D* begins, at event *q*, a journey into the internal region, as shown. As *D* crosses the horizon, event *x*, *D* receives the light emitted from *C* at event *u*—the event of *C*'s crossing the horizon. As *D* proceeds into the internal region toward the singularity, *D* receives light from *C* in the internal region. We note, however, that *D* is unable to actually see *C* strike the singularity. There is a critical point on *C*'s world-line, between u and the singularity, from which a light-pulse just manages to make it to D as D reaches the singularity. Later light-pulses emitted by C just go harmlessly into the singularity, never reaching D at all. Thus, D's visual experiences are these: As he begins to head toward the black hole, D sees C speeding up again. (Recall that, before event q, individual C was seen by D to be slowing down.) As D crosses the horizon, he sees C crossing the horizon. Thereafter, D sees C for a while in the internal region, and, finally, D gets snuffed out at the singularity, without ever having seen C get snuffed out.

It should be clear, in particular from the discussion above, that nobody will ever visually see the last moments of C's life before he reaches the singularity (except, of course, for someone who actually accompanies C). The act of "reaching the singularity" is a very personal one. The reason, of course, is that all light emitted by an individual a moment before reaching the singularity just goes directly into the singularity, and thus the light itself is snuffed out before anybody has the opportunity to experience it.

As a final example of these visual effects, let us consider the situation in which individual A, who remains in the external region, simply looks at the black hole. What will A see? Individual A will of course receive all light-pulses which reach his world-line. Such light-pulses are shown in figure 96. The pulses come from nearer and nearer the horizon, the farther back along the pulse one goes. Thus, A will see visually everybody who has ever gone into the black hole. (This is not quite true. Actually, A will see only those people who went in on A's side of the black hole. Light from those who went in on the other side, for example, would never reach A. However, A could always orbit the black hole, going over to the other side, whence he would be able to see the individuals who went in on that side.) Those who have gone in most recently will be seen the farthest from the horizon, and will appear to be moving the fastest. Those who went in a longer time ago will appear to be closer to the horizon, and will appear to be slowed down more. However, nobody who ever went into the hole will be totally invisible to individ-

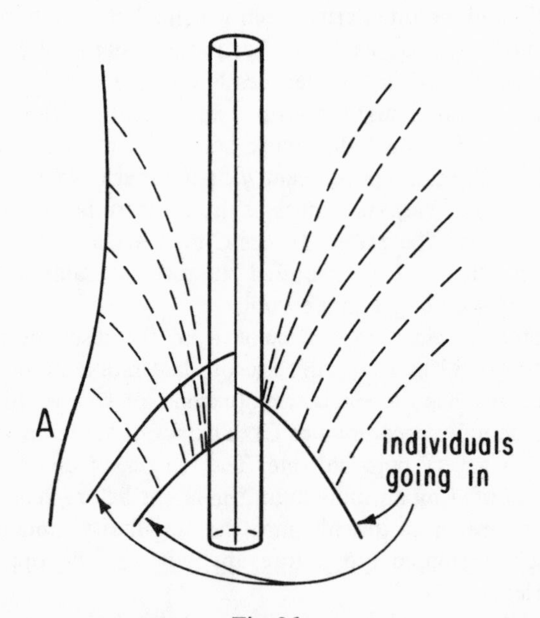

Fig. 96
A's visual impressions on looking at the black hole from the external region. *A* is able, at all times, to see every individual who has ever entered the hole.

ual *A*. The black hole thus presents a kind of "visual record" of every individual who ever made the plunge through the horizon. Only a portion of that visual record is immediately available to *A* when he is on one side of the hole. However, *A* can always go over to the other side of the hole, and recover the remainder of the record. Thus, none of the information as to who went into the black hole and who did not is ever lost to *A*.

It is now clear why a "black hole" is called what it is. *Black* refers to the fact that no light ever comes out of the hole itself, no light-pulse ever passes from the internal region to the external region. Since those in the external region receive no light from the hole itself, the hole appears black. *Hole* refers to the fact that individuals (such as *C*) can go into the object (through the horizon and into the internal region), while no individual can come back out (that is, pass from the internal region back to the external region). The horizon itself is a kind of "one-way membrane" (in but not out). Both *black* and *hole* refer to aspects of this last sentence.

We have now discussed some applications to the space-time "black hole" of two of our interpretive techniques, namely "The world-line of an individual is a timelike line" and "The world-line of a light pulse is a lightlike line." However, we have at our disposal numerous other interpretative techniques. Each of these could now be applied, to our space-time, to yield further physical predictions from the space-time. One such technique is that of figure 51, in which one makes use of the intervals to determine the elapsed times as recorded by various clocks. Using this particular technique, one could determine, for every individual's world-line we have drawn in this chapter, how many seconds that individual records as having elapsed between any two events on the world-line. Thus, we would be able to make statements of the form, "Individual *A* experienced . . . seconds between events *p* and *q* on his world-line, while *C* experienced only . . . seconds." "Individual *B* thinks that individual *C*'s clock was slowed down by a factor of . . . , while *C* thinks that *B*'s clock was speeded up by a factor of . . . ," and so on. In this way, we could in principle obtain a body of new predictions at least as extensive as those we have obtained already. Unfortunately, it is a bit awkward to carry out this particular program in detail, for the following reason. In our original diagrammatic representation of the space-time geometry (fig. 81), we showed only the light-cones, not the intervals themselves. This, recall, was done only for convenience; the cones are easy to draw and discuss, whereas the actual intervals are more difficult. The light-cones themselves, however, give one information only about "timelike related," "lightlike related," and so on—not the actual numerical value of the interval. What we would have to do, therefore, is describe in more detail, starting with the figure 81, what all the intervals are numerically. Such a description having been completed, we could embark on a detailed discussion of recorded elapsed times. Unfortunately, this description of the numerical values of the intervals, from the present viewpoint, would be rather lengthy. Since, furthermore, the physical information which would flow from such a description is not all that interesting, we omit this statement of the intervals in more detail. We may, however, just summarize a few of the physical conclusions which would stem from such a program. In figure 83, B would experience a smaller elapsed time between p and q than A would. In figure 84, C would experience a smaller elapsed time between u and the singularity by taking the railroad-track world-line than by taking the "direct route." In figure 85, C, by delaying the start of his return journey to A until he is very near event u, can cause his apparent elapsed time in the return journey to A to be as small as he wishes, even though A may regard C as having been away for a very long time (say, years). Thus, in this example, C might return to A 100 years later (according to A), even though C might have aged only a few days.

As a final example of application to this space-time geometry of interpretative techniques, let us consider the following. "The world-surface of a rope is a two-dimensional surface." The experimental situation we wish to consider is this. As usual, A remains outside the hole, while C ventures in. This time, however, C first ties one end of a long rope around his waist, and A is instructed to feed the rope out as C goes into the hole. A space-time diagram for this experiment is shown in figure 97. For convenience of representation, let marks have been made at regular intervals along the rope. The world-lines of these marks are also shown in the diagram. Note that "new mark world-lines continually emerge from A," the geometrical statement of the physical idea that A is feeding out the rope. The world-lines of the marks must be timelike. Thus, in particular, those lines in the internal region must proceed into the singularity. In physical terms, after C reaches the singularity his end of the rope is continually pulled into the singularity (that is, successive marks become snuffed out).

Now let us consider the following situation. At event q, individual A begins to become a bit worried about C, and decides to try to pull him out of the hole using the rope. So, A pulls on

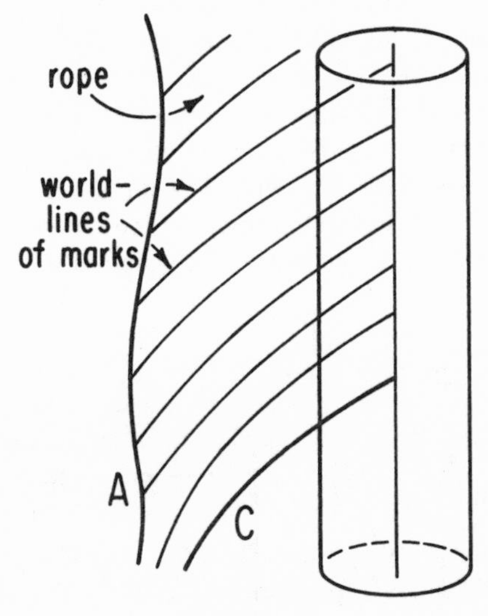

Fig. 97
Observer *A* lowers *C* into the hole by means of a rope, and continues to feed out the rope at all later times. Included in the figure are the world-lines of "marks" made on the rope. The rope, then, is continually being fed into the singularity.

the rope at event *q* (fig. 98). After event *q*, "mark world-lines" no longer emerge from *A*, but rather begin to come back to *A*, as *A* pulls in the rope. Now we know that the "mark world-lines" from the internal region will never come back to the external region (since these are timelike lines). The net result of *A*'s pulling on the rope, then, is that there will be fewer marks available between *A* and the hole. Physically, this means that the rope is being stretched. In particular, then, *A* will have to pull harder and harder on the rope in order to continue reeling it in (that is, in order to continue to receive on his world-line additional marks). Eventually, presumably, the rope will break, as illustrated in figure 98. One end will just continue falling into the black hole, eventually being absorbed by the singularity, while *A* will be able to recover the rest, all of which is illustrated in the diagram. The upshot of all this is that *A* will not be successful in "pulling C out of the black hole using the rope." Of course none of this is surprising, since we saw before that C, once he has crossed the horizon, will never be able to return to the external region. What we do see in this example is what actually happens physically under a particular physical setup where A does try to bring C out. Also in figure 98 is event w, at which C (naively, we now know) decides to signal A that he would like to be pulled out now. The signal is given by shaking

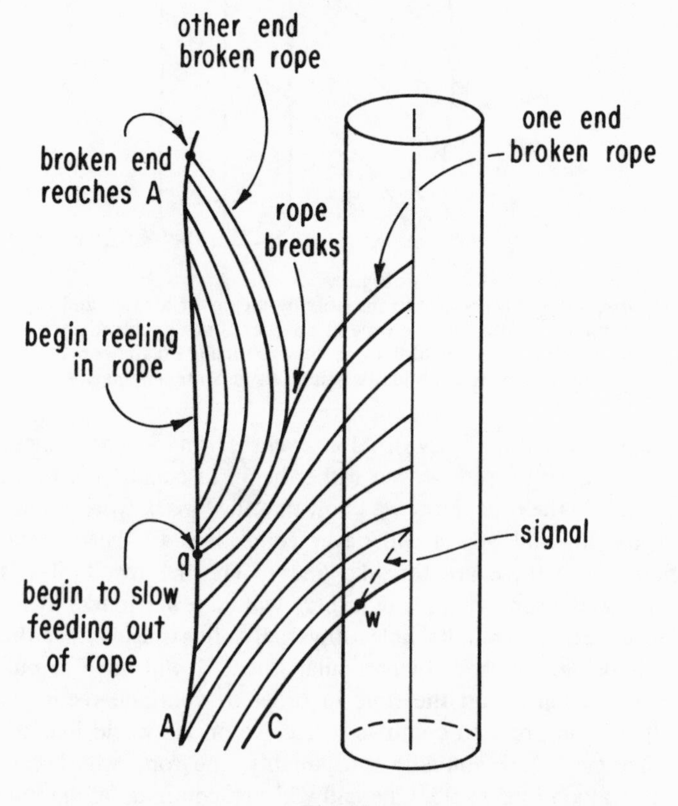

Fig. 98

A space-time diagram of what happens if A decides to retrieve C by pulling on the rope. Rather than C's returning, the rope breaks.

the rope. Unfortunately, this signal must travel along a timelike world-line. Since, furthermore, this world-line originates in the internal region, the world-line goes harmlessly into the singularity. Thus, *C*'s signal for help never reaches *A*.

This completes our discussion of various physical phenomena which arise from the space-time geometry labeled *black hole*. We emphasize both the diversity and detail of these physical predictions. For the techniques discussed here (even for "The world-line of an individual is timelike; the world-line of a light pulse is lightlike.") we have described only a small fraction of the many possible physical predictions that could be made. Furthermore, we have dealt with only a few of the many interpretative techniques that are available. Finally, there are numerous additional interpretative techniques that are not even mentioned here. (One example of such unmentioned techniques: What are the intensities of the received light-pulses?) Putting all these together, it would be very easy to go on for several hundred more pages making more and more physical predictions about this space-time. We could invent complicated combinations with individuals, clocks, light-pulses, ropes, sheets of paper, radiation; asking about elapsed times, intensities and frequencies of received light-pulses, tensions of ropes, and so on. This whole constellation of physical phenomena, of which we have here given a very small fraction, goes under the name "physical phenomena associated with the space-time geometry 'black hole.' "

Recall how the theory works. One finds models—space-time geometries, examines the associated physical phenomena, and then asks whether or not such phenomena are found in our world. By this activity, one hopes to "converge" on that model which might represent our world. We now wish to get a glimpse of how this program works in action. We have in our possession one particular model, "black hole," and we have at least some of the physical phenomena associated with it. The next question, then, is this: Do we, or do we not, see in our world some body of physical phenomena rather like that we have been discussing associated with "black hole"? If not, then one would reject this space-time geometry, and move on to another. It turns out, however, that we do, at least in some sense, see such phenomena in our world. A brief summary of the situation follows.

In figure 27, we briefly mentioned double-star systems. Such systems are of course actually observed in our world. One looks up into the sky, and sees two nearby stars. As time goes on, the two points of light in the sky appear to describe circles about each other. One sees the stars orbiting each other. Occasionally, however, one sees the following phenomenon in the sky. One sees but a single point of light, but, over time, that point appears to describe a small circle in the sky. What could this be? It is very difficult to see, from the physics we think we understand, how a star could be going around in circles out there on its own initiative. A more likely explanation is that this is essentially a double-star system, but, for some reason, we do not see the "other star." What could this other object be? Again, we try first for the simplest explanations. It could very well be just a very dim star—so dim that we do not see it. (There are both theoretical and experimental reasons for believing that such dim stars exist in the universe.) This explanation, however, runs into several difficulties. First, one can determine the mass of the unseen object, a determination which is carried out by watching the orbit of the seen star and computing how much unseen mass would be required to yield such an orbit. In many cases, the unseen object turns out to have a rather large mass—so large in fact that our present understanding of stellar structure would suggest that one cannot have very dim stars of these large masses. This observation, then, suggests that our object is not a dim star. Further evidence arises from the fact that one receives from such systems not only light but also X rays. How do the X rays originate? It turns out that ordinary stars—and even ordinary double-star systems—do not normally produce X rays. These X rays could be accounted for if, somewhere in this system, there is a gas of electrons compressed to a rather high density. Now, if our system is to be an ordinary double-star system, consisting of one regular star and one rather dim star, it is difficult to see where in that system there could be such a compressed gas of electrons. Suppose, however, that the "dim star" was in fact a black hole, and that electrons, escaping from the surface of the seen star, fall into the hole. As they fall in, one calculates, the electrons would be pushed closer together. The result—a compressed gas of electrons. (This is, of course, part of the physical phenomena associated with the space-time geometry "black hole.") It seems possible, therefore, that the unseen object is actually a black hole (therefore, "unseen"), and that the X rays result from these electrons falling into the hole.

The above is the thrust of the argument that black holes have been observed. It sounds tenuous because it is. It is perfectly possible that the argument about the mass is wrong because we do not understand stellar structure well enough, or that the argument about X rays is wrong because there is some other mechanism for the production of X rays that we haven't thought of. I would regard it as by no means out of the question that, ten years from now, it will be widely felt that black holes have never been observed. Nonetheless, one does the best one can with what evidence one can gather, and the best one can do at present to "explain" these various observations is the explanation: the space-time geometry "black hole." As time goes on, new evidence will undoubtedly come in, and the tide of opinion will shift one way or another.

Conclusion

This, then, is Einstein's theory of relativity. Its distinctive feature, perhaps, is the introduction of the idea of space-time—the idea that everything of physical interest in the world, including oneself, is to be set down therein once and for all. In the Galilean view, space-time is a luxury; in relativity, a necessity. The theory has a strongly operational flavor; one does not easily assign attributes to objects in the world except insofar as information from those objects reaches one and is interpreted according to one's own devices. Indeed, one of the charms of the theory is that it illuminates a peculiar and delicate kind of interaction between oneself and the rest of the world. In one sense, the world is just "there" in space-time, with no need of the observer; in another, it reduces almost to a moving picture that one watches. I hope that we have made a case that this theory is neither strange nor difficult. Indeed, looking back from the vantage point of the theory, its message—introduce events, assemble into space-time, represent and describe physics within this framework—seems almost tautological. It is as though the theory is trying to be simple and natural.

The general theory of relativity is currently of very little use in building an airplane or solving the energy crisis. Very likely, it never will be. Then what good is it? Man has at least two sides to his nature: his physical needs and his intellectual life. Part of the latter consists of acquiring as deep an understanding as he can of the nature of the physical world in which he lives.

This activity has gone on for centuries, and this activity will presumably continue. Let it be granted then that the search for understanding as an end in itself is a viable human activity. It is in this general area that general relativity largely falls today. Of course, we have no guarantee that it will always remain there. It is altogether possible that, at some time in the future, material benefits will flow from some application of the theory. Nonetheless, I would personally regard the chances for such benefits, at least in my lifetime, as rather remote, and would therefore not regard the expectation of such benefits as the primary source of interest in the theory. I find the theory interesting because through it I feel that I understand nature better.

Our entire discussion of general relativity, from its basic ideas to its more technical aspects, has really been a mere excuse for something else. That "something else" is a statement of how physics operates (or, what is the same thing, of how physicists operate). I hope that I have managed to convey the feeling that thinking about the physical world is not in any essential way different from thinking about anything else. Physics is a human activity, and, as such, it is vague, uncertain, and judgmental. I would claim that physicists as a group are no more logical, certain of their conclusions, or conversant with "truth" than anybody else. This is a circumstance not to regret but to relish.

Index